Darkening Air:
The Invisible Threat:

Dr Ajay Kumar Gupta
Pritanshu

DEPARTMENT OF BIOTECHNOLOGY
MAHARISHI MARKANDESHWAR UNIVERSITY
MULLANA- 133 207 (HARYANA) INDIA

ACKNOWLEDGEMENTS

All works and researches done by worthy scientists are dully cited and have formally referred in this book. Should any authors have any concern are most welcome to contact the authors. We shall try to rectify the same with utmost care and obligations.

Dr. Ajay Kumar Gupta
Pritanshu

Preface

The book is titled as "" as it answers all the basic questions arises in mind regarding air pollution. With the advent of technology we have invited many problems. The deterioration of air is one of the several aftermaths. It has been discussed for a long time but to no avail as problems due to air pollution are increasing day by day. The situation is horrendous in developing countries. This book discusses all the aspects of encumbering air pollution including its causes and effects. The book explains each and every topic in simple and easy language. The book includes empirical data of various eminent organizations .It also embodies indoor pollution which is gaining its presence very fast.

The basic aim of this book is to provide comprehensive, rigorous, and balanced introduction to air pollution and its effect on environment and living beings. It is a guide to learn a critically important and sometimes difficult subject. In writing this book, we had in mind its usefulness for undergraduate and post graduate students in the Environmental Sciences. Authors' main focus was on to the point and brief discussion of the topics and hence made a successful handbook of the title.

Dr. A.K. Gupta
Pritanshu

Contents

Chapters	Page Nos.
1. Introduction	1-13
2. Sources of pollution	14-36
3. Types of Pollutants	37-81
4. Effect of pollution	82-205
References	206-225

Seven sins: wealth without work, pleasure without conscience, commerce without morality, worship without sacrifice, politics without principle, knowledge without character and **science without humanity.**

-- Mahatma Gandhi

1. Introduction

"There's so much pollution in the air now that if it weren't for our lungs there'd be no place to put it all." ---*Robert Orben quotes (US magician and comedy writer, b. 1927)*

Very first resource human utilize ,even unknowingly , as soon as he takes birth on this planet is AIR. Being on earth we our endowed with natural resources and air is one of them. Air provide us with oxygen which is essential for our body cells to function. Body can live without water and food for few days but cannot survive for even a few seconds without air. For this survival quality of air also matters .Not only humans but all living creatures consume air for their survival. Always proud to be Ingenious creature among all , how much justice have we done with our environment ? We can't refuse that planet is nurturing us and we are neglecting it. It is miserable that we Indians who are known to hold high reverence for plants, animals and natural resources have not been successful to take care of them. Our condemning is not sufficient, we all our responsible for maintenance of our atmosphere so we all should now stop being ignorant.

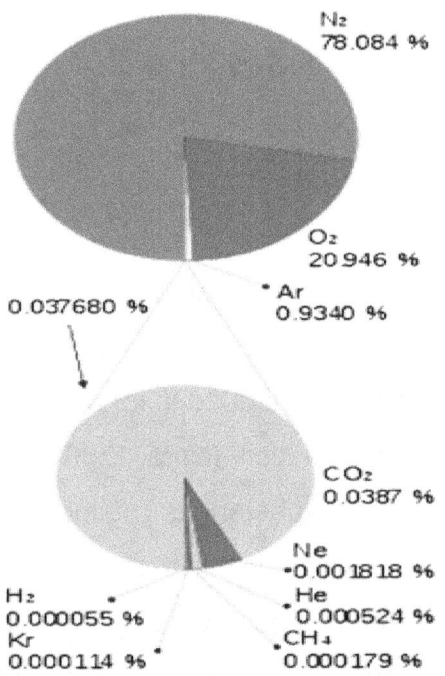

Figure 1.1: Composition of air(CPCB.2012).

Table 1.1: The gaseous composition of unpolluted air. (O.P Singh, Air Pollution: Types, Sources and Abatement)

The Gases	Parts per million(vol)
Nitrogen	756,500
Oxygen	202,900
Water	31,200
Argon	9,000
Carbon Dioxide	305
Neon	17.4
Helium	5.0
Methane	0.97-1.16
Krypton	0.97
Nitrous oxide	0.49
Hydrogen	0.49
Xenon	0.08
Organic vapours	ca.0.02

Generally, air is mainly composed of nitrogen and oxygen (99% by volume) and other gases including water vapor contribute to about 1%. Nitrogen, oxygen, water vapor and inert gases constitute 99.95of air. But the present-day atmosphere is quite different from the natural atmosphere that existed before the Industrial Revolution (circa 1760 [T.S. Ashton,1948]), in terms of chemical composition. It is because of substances (pollutants) released into the air due to both manmade and natural processes such as volcanic activities, dust storms, forest fire etc. Also pollutants have been observed to persist in the environment, to be capable of long-range transport, bioaccumulate in human and animal tissue, biomagnifies in food chains, and to have potential significant impacts on human health and the environment. In recent years the air pollutants has been aggravated by a number of reasons such as the population growth, urbanization, rapid economic development, industrialization and increasing traffic and levels of energy consumption. Pollutants can be classified as primary or secondary pollutants(Daly, A. and P. Zannetti. 2007). Primary pollutants are substances that are directly emitted into the atmosphere from sources as number of primary pollutants resulting from combustion of fuels and industrial operations .The main primary pollutants known to cause harm in high enough concentrations are carbon compounds, such as CO, CO_2, CH_4, and volatile organic compounds (VOCs) , nitrogen compounds, such as NO, N_2O, and NH_3, sulfur compounds, such as H_2S and SO_2, halogen compounds, such as chlorides, fluorides, and bromides and particulate matter (PM or "aerosols"), either in solid or liquid form.

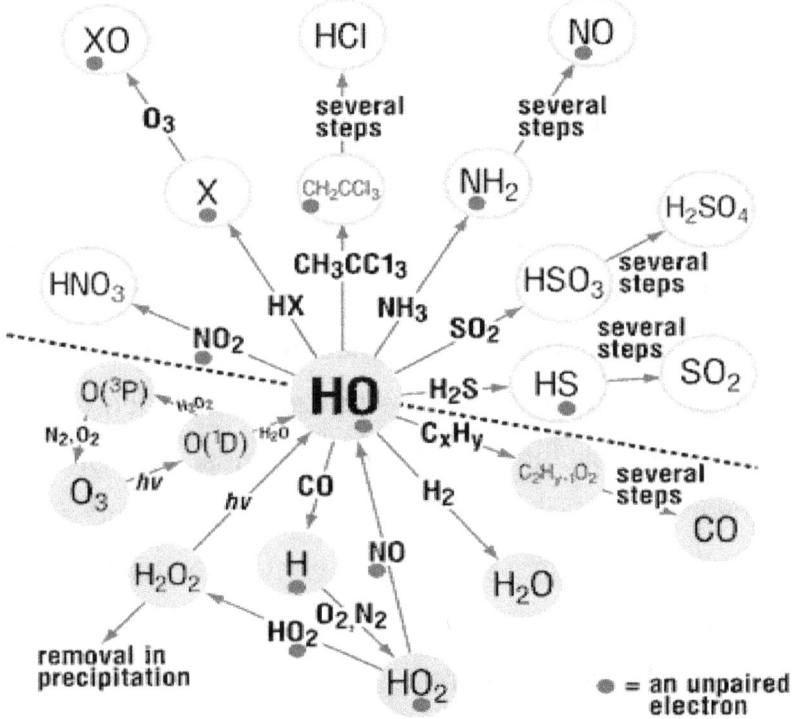

Figure 1.2: Chemical reactions in the atmosphere .(Madronich S. Chemical reactions in the atmosphere, Lecture 32, used to be available on North Arizona University's website but is no longer accessible.)

Whereas secondary pollutants are not directly emitted from sources, but instead produced due to reaction of primary pollutants(also called "precursors") in the atmosphere , The main secondary pollutants known to cause harm in high enough concentrations are NO_2 and HNO_3 formed from NO, ozone (O_3) formed from photochemical reactions of nitrogen oxides and VOCs, sulfuric acid droplets formed from SO_2 , nitric acid droplets formed from NO_2, sulfates and nitrates aerosols (e.g., ammonium (bi)sulfate and ammonium nitrate) formed from reactions of sulfuric acid droplets and nitric acid droplets with NH3, respectively and organic aerosols formed from VOCs in gas-to-particle reactions. In the 20th century, it was recognized that petroleum products are responsible for a new type of "smog", a photochemical summertime smog composed of secondary pollutants such as ozone(Daly, A. and P. Zannetti. 2007).Photochemical smog (http://www.epa.qld.gov.au/environmental_management/air/air_quality_monitoring/air_pollutants/ozone/) was first recognized (http://www.pbs.org/now/science/smog.html) in the city of Los Angeles in the 1940s. After decades of research, the smog was identified as the product of photochemical reactions involving "precursors (nitrogen oxides and VOC) and sunlight, with the production of ozone and other secondary chemicals. While nitrogen oxides are emitted by a wide variety of sources, automobiles mostly emit VOCs, even though contributions can be found from vegetation and common human activities, such as bakeries (The by-products of fermenting yeast are CO2 and ethanol (a VOC). So baking bread in large quantities can contribute to ozone formation due to emissions of VOCs.). Some secondary pollutants – sulfates, nitrates, and organic particles – can be transported over large distances, such as hundreds and even thousands of miles. Wet and dry deposition of these pollutants contributes to the "acid deposition"problem (often called "acid rain")(http://en.wikipedia.org/wiki/Acid_rain), with possible damage to soils, vegetation, and susceptible lakes.

As air is present everywhere so is the pollutants in it.Pollution is everywhere both indoor and outdoor places. Consequently, causing premature deaths of humans and damaging flora and fauna, transport system and our cultural assets (monuments and historical buildings).

Aphorism of Air Pollution

"The presence in the atmosphere of one or more contaminants in such quantity and for such duration as is injurious, or tends to be injurious, to human health or welfare, animal or plant life."

(O.P Singh, Air Pollution: Types, Sources and Abatement)

Simillarly, environmental pollution is the contamination of the ecosystem that causes instability, disorder, harm or discomfort to the physical systems or living organisms. Environmental factors have important links with infectious as well as non-infectious diseases of both acute and chronic nature. Global burden of disease attributable to selected sources of environment like water sanitation and hygiene, urban outdoor and indoor pollution, occupational carcinogens, noise and airborne particulates has been assesses to be 8-9 %, measured either in terms of mortality or 'disability adjusted life years' (DALYs)(Kavita Gulati et all ,2010).

People are exposed to toxic air pollutants in many ways that can pose health risks, such as by:

- Breathing contaminated air.
- Eating contaminated food products, such as fish from contaminated waters; meat, milk, or eggs from animals that fed on contaminated plants; and fruits and vegetables grown in contaminated soil on which air toxics have been deposited.
- Drinking water contaminated by toxic air pollutants.
- Ingesting contaminated soil. Young children are especially vulnerable because they often ingest soil from their hands or from objects they place in their mouths.
- Touching (making skin contact with) contaminated soil, dust, or water (for example, during recreational use of contaminated water bodies).

Once toxic air pollutants enter the body, some persistent toxic air pollutants accumulate in body tissues. Predators typically accumulate even greater pollutant concentrations than their contaminated prey. As a result, people and other animals at the top of the food chain that eat contaminated fish or meat are exposed to concentrations that are much higher than the concentrations in the water, air, or soil (http://www.epa.gov/air/toxicair/newtoxics.html).

Air pollution aggravated by human per se is among biggest dangers to our planet, metamorphosing nourishing environment to killer. A pivotal study linking air particulates to adverse health outcomes was published by a group of investigators from Harvard in which they showed that people living in cites with high levels of particulates had lower life expectancy than people living in more polluted urban areas that had subordinate levels of particulates. Their comparison was based on data collected from six cities across the United States. (Dockery et al.1993).Situation is rather grave, every other day scientists are finding its worrisome effects on living beings. Data collected through satellites clearly show heavy amount of dangerous air-polluting emissions and their transboundary movement. Scientists and researchers all around the world are actively engaged in identifying hotspots, analyzing trends, apply best solutions and monitoring the effectiveness of those mitigation efforts.

There are many environmental issues in India. Air pollution is one of those challenges. According to data collection and environment assessment studies of World Bank experts, between 1995 through 2010, India has made one of the fastest progresses in the industrial world, but least in addressing its environmental issues and improving its environmental quality. Thus, India has a long way to go to reach environmental quality similar to those enjoyed in developed economies. Pollution remains a major challenge and opportunity for India.

Major sources of pollutants in Indian air is fuelwood and biomass burning, vehicle emission (Atmanand et al. 2009) (The World Bank.2002) and industrial exhaust. Rapid increase in urban population brought with it lots of problems such as unplanned urban development, increase in consumption patterns and higher demands for transport, energy and other infrastructure. The problem of air pollution in urban areas is also aggravated due to inadequate power supply for industrial, commercial and residential activities due to, which consumers have to use diesel-based captive power generation units emitting high levels NO_X and SO_X. In addition, non-point sources such as waste burning, construction activities, and roadside air borne dust due to vehicular movement also contribute to the total emission load (UNEP "India; State of the Environment 2001").

There are also some other responsible factors for atmospheric deterioration in India.Vehicle and industrial emissions are worsened by fuel adulteration and poor fuel combustion efficiencies from traffic congestion and low density of quality.Moreover, it has been noticed that in developing countries, less attention is paid to environmental protection, environmental standards are often inappropriate or not effectively implemented, and pollution control techniques are not yet fully developed. In order to gain profit at the expense of nature some Indian taxis, trucks and auto-rickshaws run on adulterated fuel blends. As per World Bank adulteration of gasoline and diesel with lower-priced fuels is common in South Asia, including India (Atmanand et al. 2009). As a result of adulteration amount of emissions of harmful pollutants increase and further add to urban air pollution. Financial incentives arising from differential taxes are generally the primary cause of fuel adulteration. In India and other developing countries, gasoline carries a much higher tax than diesel, which in turn is taxed more than kerosene meant as a cooking fuel, while some solvents and lubricants carry little or no tax.As fuel prices rise, the public transport driver cuts costs by blending the cheaper hydrocarbon into highly taxed hydrocarbon. The blending may be as much as 20-30 percent. For a low wage driver, the adulteration can yield short term savings that are significant over the month. In doing so safety and prosperity of health, environment and ultimately of life is put on stake. At the same time the reduced life of vehicle engine and higher maintenance costs is not taken into account. Adulterated fuel increases tailpipe emissions of hydrocarbons (HC), carbon monoxide (CO), oxides of nitrogen (NOx) and particulate matter (PM). Air toxin emissions — which fall into the category of unregulated emissions — of primary concern are benzene and polyaromatic hydrocarbons (PAHs), both well known carcinogens. Kerosene is more difficult to burn than gasoline; its addition results in higher levels of HC, CO and PM emissions even from catalyst-equipped cars. The higher sulfur level of kerosene is another issue. Fuel adulteration is essentially an unintended consequence of tax policies, inflation and the attempt to control fuel prices, in the name of need. What so ever, poor air quality and deteriorating environment is the ultimate consequence. This problem is not unique to India, but prevalent in many developing countries including those outside of south Asia (http://www.ilo.org/oshenc/part-vii/environmental-health-hazards/item/497-industrial-pollution-in-developing-countries). This problem is largely absent in economies that do not regulate the ability of fuel producers to innovate or price based on market demand.

Previously limited to metropolitan cities, traffic congestion is now disseminated to even small cities and towns. Traffic congestion is caused for several reasons, like increase in number of vehicles per kilometer of available road, insufficient road infrastructure as lack of intra-city divided-lane highways and intra-city expressways networks, lack of inter-city expressways, traffic accidents and chaos from poor enforcement of traffic laws. Situation is so poor that vehicles stuck for hours in traffic. Traffic gridlock in Delhi and other India cities is extreme. (India Today. September 5, 2010). The average trip speed on many Indian city roads is less than 20 kilometers per hour; a 10 kilometer distance can take 30 minutes, or more. At such speeds, vehicles in India emit air pollutants 4 to 8 times more than they would with less traffic congestion; Indian vehicles also consume a lot more carbon footprint fuel per trip, than they would if the traffic congestion was less.Traffic congestion reduces average traffic speed. At low speeds, scientific studies reveal, vehicles burn fuel inefficiently and pollute more per trip. For example, a study in the United States found that for the same trip, cars consumed more fuel and polluted more if the traffic was congested, than when traffic flowed freely. At average trip speeds between 20 to 40 kilometers per hour, the cars pollutant emission was twice as much as when the average speed was 55 to 75 kilometers per hour. At average trip speeds between 5 to 20 kilometers per hour, the cars pollutant emissions were 4 to 8 times as much as when the average speed was 55 to 70 kilometers per hour (Matthew Barth et al. 2009). Fuel efficiencies similarly were much worse with traffic congestion.

Fig 1.3: Showing Particulate pollutant levels far above NAAQS in one of the metropolitan city of India, Kolkata. (Manas Ranjan Ray and Twisha Lahiri .2010." Air Pollution and its Effects on Health –Case Studies , India" CNCI,Kolkata)

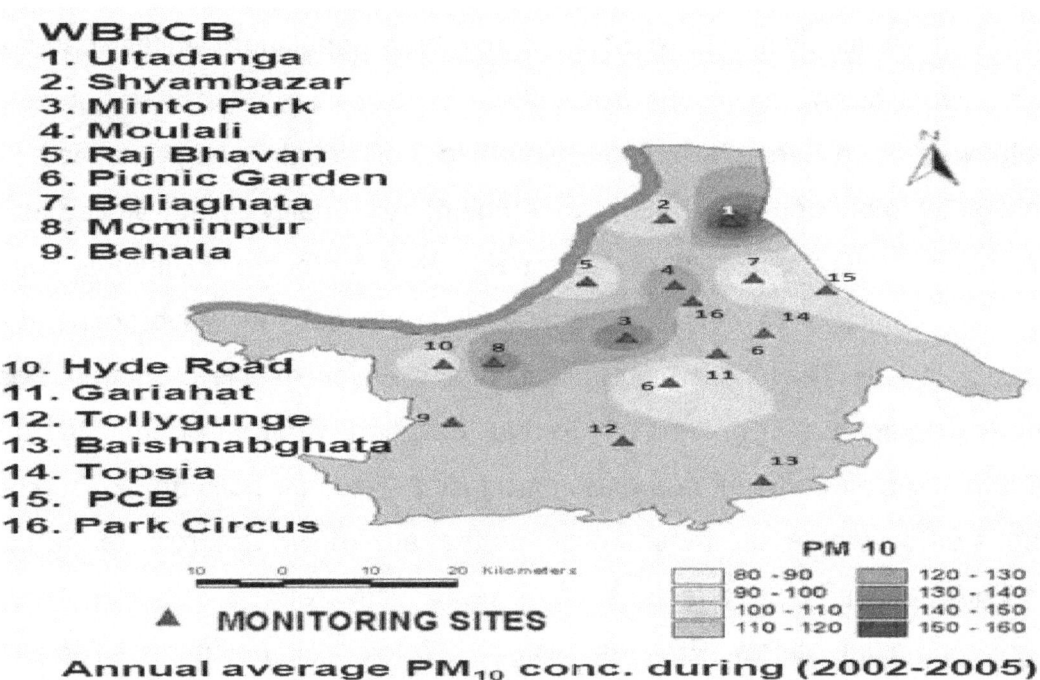

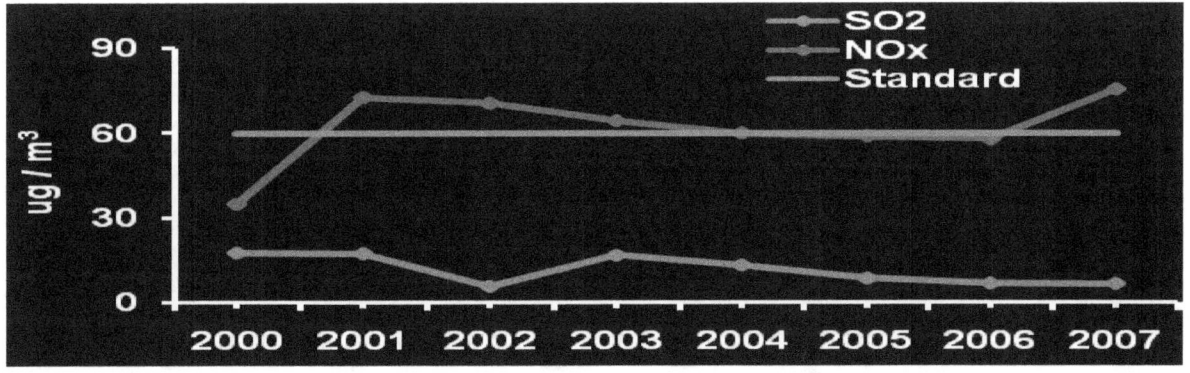

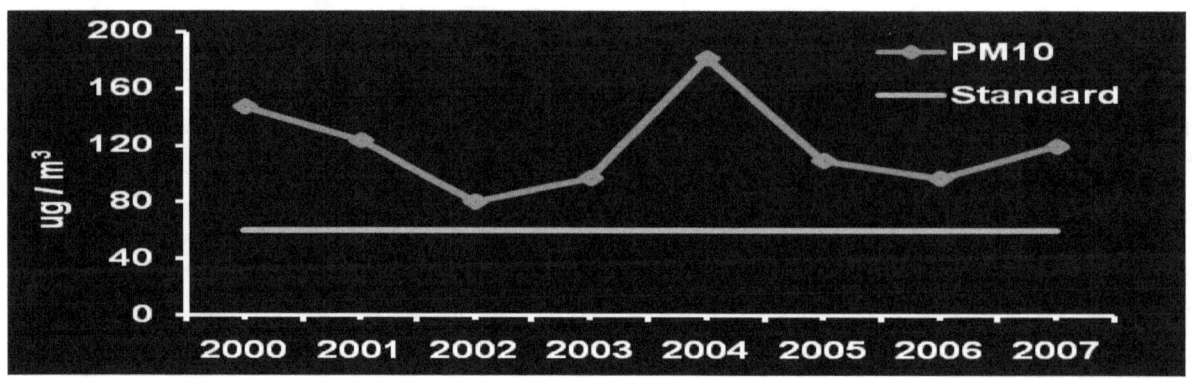

So, the rapid and seemingly unending increase in vehicle population, limited road space and carrying capacity and emissions from different intersectorial activities, burning of agricultural waste in remote areas ,burning of coal and biomass and study/investigation/monitoring of the Urban Ambient Air Quality is not limited to few parameters such as NOx, SO2, CO etc. The study of hazardous air pollutants such as ozone, volatile organic compounds (VOCs), ketones and aldehydes have acquired new dimensions for obvious reasons, which include their adverse effects on human health, vegetation (ex: crops) and materials(Central Pollution Control Board ,2010).The additional term toxic air pollutant has been introduced .Toxic air pollutants, also known as hazardous air pollutants, are those pollutants that are known or suspected to cause cancer or other serious health effects, such as reproductive effects or birth defects, or adverse environmental effects. EPA has listed 187 toxic air pollutants (Table 1.2). Examples of toxic air pollutants include benzene, which is found in gasoline; perchloroethylene, which is emitted from some dry cleaning facilities; and methylene chloride, which is used as a solvent and paint stripper by a number of industries. Examples of other listed air toxics include dioxin, asbestos, toluene, and metals such as cadmium, mercury, chromium, and lead compounds (http://www.epa.gov/air/toxicair/newtoxics.html).

The findings of source apportionment studies conducted by CPCB have revealed that the release of PM10 from the transport sector is 20.5% in Delhi and as high as 48.3% in Chennai. The other major sources of PM10 emissions in Indian urban environment are garbage burning, road dust, DG sets, construction activities etc. The CPCB has also carried out a study of the urban air quality in Kolkata for source identification and estimation of ozone, carbonyls, NOx and VOC emissions. The study details and its findings which have been published by CPCB in its document Control of Urban Pollution Series CUPS/72/2010-11, include cigarette smoke as one of the source of VOC emissions in urban ambient air. The keeping of the Ambient Air Quality from the point of sustainability not only requires ensuring of a sustainable growth of the vehicle population and sustainable siting of developmental activities but also the sustainability of the use of fuels for various purposes including as automotive fuel. The sustainability of automotive fuels includes the type, affects and acceptability of the HAPs that are released from different fuel based vehicles (Central Pollution Control Board ,2010).

The government has taken a number of measures such as legislation, emission standards for industries, guidelines for siting of industries, environmental audit, EIA, vehicular pollution control measures, pollution prevention technologies, action plan for problem areas, development of environmental standards, and promotion of environmental awareness(UNEP "India; State of the Environment 2001"). Emission standards express the allowable concentrations of a contaminant at the point of discharge before any mixing with the surrounding air.

Table 1.3: Revised National Ambient Air Quality Standards (NAAQS).[NAAQS Notification dated 18th November, 2009]. (CPCB.2012)

S.No.	Pollutants	Time Weighted Average	Concentration in Ambient Air	
			Industrial, Residential, Rural and other Areas	Ecologically Sensitive Area (notified by Central Government)
1	Sulphur Dioxide (SO_2), µg/m$_3$	Annual*	50	20
		24 Hours**	80	80
2	Nitrogen Dioxide (NO_2), µg/m$_3$	Annual*	40	30
		24 Hours**	80	80

3	Particulate Matter (Size <10μm) or PM$_{10}$ μg/m$_3$	Annual*	60	60
		24 Hours**	100	100
4	Particulate Matter (Size <2.5 μm) or PM$_{2.5}$ μg/m$_3$	Annual*	40	40
		24 Hours**	60	60
5	Ozone (O$_3$), μg/m$_3$	8 hours**	100	100
		1 hours **	180	180
6	Lead (Pb), μg/m$_3$	Annual*	0.50	0.50
		24 Hour**	1.0	1.0
7	Carbon Monoxide (CO), mg/m$_3$	8 hours**	02	02
		1 hours **	04	04
8	Ammonia (NH$_3$), μg/m$_3$	Annual*	100	100
		24 Hour**	400	400
9	Benzene (C$_6$H$_6$), μg/m$_3$	Annual*	05	05
10	Benzo(a)Pyrene (BaP)- particulate phase only, ng/m$_3$	Annual*	01	01
11	Arsenic (As), ng/m$_3$	Annual*	06	06
12	Nickel (Ni), ng/m$_3$	Annual*	20	20

Annual Arithmetic mean of minimum 104 measurements in a year at a particular site taken twice a week 24 hourly at uniform interval.
** 24 hourly 08 hourly or 01 hourly monitored values, as applicable shall be complied with 98% of the time in a year. 2% of the time, they may exceed the limits but not on two consecutive days of monitoring.

The Air (Prevention and Control of Pollution) Act was passed in 1981 to regulate air pollution and there have been some measurable improvements. There have been success reports such as the reduction of ambient lead levels (due to introduction of unleaded petrol) and comparatively lower SO2 levels (due to progressive reduction of sulphur content in fuel) (UNEP "India; State of the Environment 2001")(CPCB.2012). However, despite all these measures, air pollution still remains one of the major environmental problems. The 2012 Environmental Performance Index ranked India as having the poorest relative air quality out of 132 countries (Yale University. 2012).According to Dr Sachin Ghude of the Indian Institute of Tropical Meteorology (IITM), rapid industrialization, urbanization and traffic growth are most likely responsible for the increase.Because of varying consumption patterns and growth rates,the distribution of emissions vary widely across India.In order to mitigate the causes of pollution,policy makers need to know the hardest hit regions(© Phys.org™ 2003-2013). India has low per capita emissions of greenhouse gases but the country as a whole is the third largest after China and the United States. (International Energy Agency, France.2011).

2.Sources of pollutants

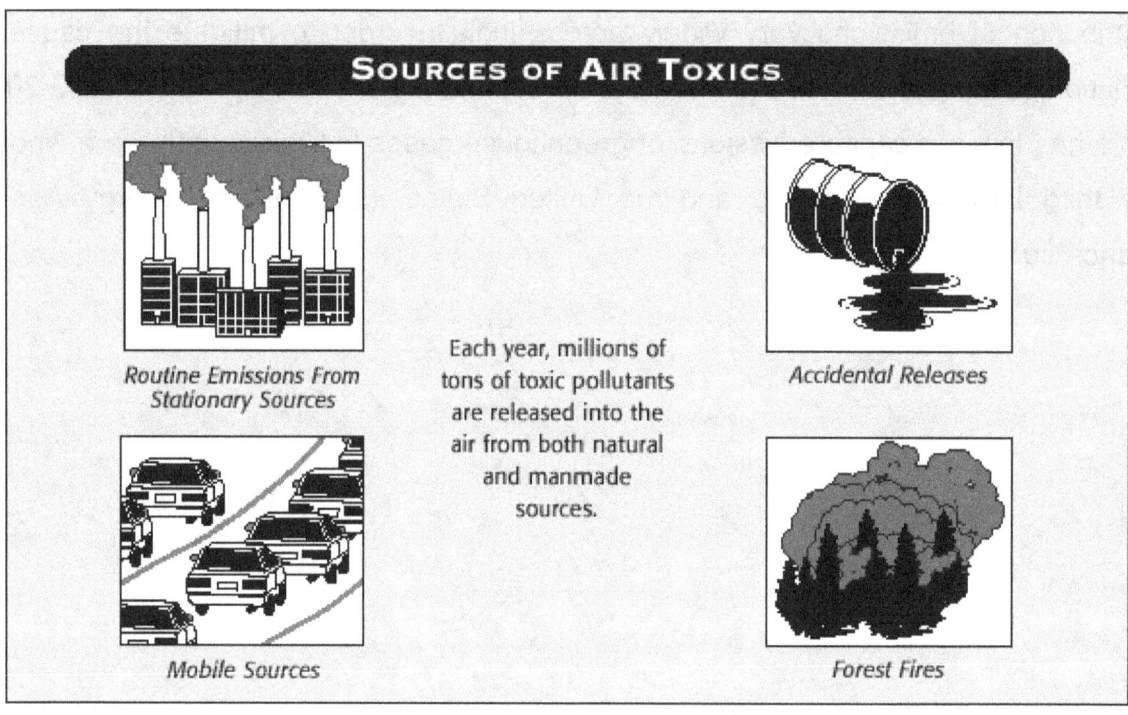

Fig 2.1: Common sources of air toxics. (www.epa.org)

Sources of air pollution refer to the various locations, activities or factors which are responsible for the releasing of pollutants into the atmosphere. It is evident that that these sources are either man made or natural .So two major categories to which whole sources belongs to are anthropogenic sources and natural sources. Most air toxics originate from human-made sources, including mobile sources (e.g., cars, trucks, buses) and stationary sources (e.g., factories, refineries, power plants), as well as indoor sources (e.g., some building materials and cleaning solvents). Some air toxics are also released from natural sources such as volcanic eruptions and forest fires (http://www.epa.gov/air/toxicair/newtoxics.html).

Fig 2.2 : Classification of sources of Pollutants.

Anthropogenic sources (man-made sources) include complete range of sources of pollutants human is responsible for. These are mostly related to burning different kinds of fuel. Power plants, traditional biomass burning,motor vehicles, marine vessels, aircraft ,chemicals, dust and controlled burn practices in agriculture and forestry management. Waste deposition in landfills, nuclear weapons, toxic gases, germ warfare and rocketry , Fumes from paint, hair spray, varnish, aerosol sprays and other solvents are all fall in this division .Anthropogenic sources can be further divided into three classes mobile sources, stationary sources and area sources. The category labeled mobile sources includes vehicles and pieces of equipment that by design and function emit pollutants while driving or moving. Examples of this source type include: On-Road vehicles (cars, motorcycles, all types of trucks and buses); and Non-Road equipment and vehicles (tractors, dozers, locomotives, boats, lawnmowers, and mobile generators). Stationary Sources includes air pollution sources that do not move from location to location as with the Mobile Sources group. Examples of some Stationary sources are: chemical facilities, fuel terminals, power generating facilities, manufacturing facilities, and label printers. The category labeled Area Sources is much broader by definition and includes several smaller sized Stationary Sources that emit a similar pollutant(s) and are clustered together in one geographical area. Examples of some Area Sources are: parking decks, dry cleaners, gas stations, and automotive refinishing/painting operations (http://airquality.charmeck.org).

Second major category embodies all natural sources of pollutants. There are varieties of natural causes, not all of which are within human control. e.g. dust from natural sources, Methane, emitted by the digestion of food by animals, for example cattle Radon gas from radioactive decay within the Earth's crust. It is considered to be a health hazard. Radon gas from natural sources can accumulate in buildings, especially in confined areas such as the basement and it is the second most frequent cause of lung cancer, after cigarette smoking. Smoke and carbon monoxide are emitted from wildfires. Vegetation, in some regions, emits environmentally significant amounts of VOCs on warmer days. These VOCs react with primary anthropogenic pollutants—specifically, NO_x, SO_2, and anthropogenic organic carbon compounds—to produce a seasonal haze of secondary pollutants. Volcanic activity, produce sulfur, chlorine, and ash particulates. Active volcano pours great amounts of ash and toxic fumes into the atmosphere and lead to the deterioration of air quality (O.P Singh, Air Pollution: Types, Sources and Abatement).

Using the WHO environmental health criteria for assessing human exposure as a guideline, exposure is defined as "contact over time and space between a person and one or more biological, chemical and physical agents" (MacIntosh, 2000). As to transport-related air pollution, the most important aspects of this exposure include:

• concentration, in µg/m3 or another equally valid metric;

• duration in seconds, minutes, hours, days, weeks, months and years;

•settings, such as location of residence and workplace and transport mode (for example, commuting); and

•exposed population, whether this comprises the general population, subgroups and/or individuals.

We cannot turn a blind eye to the fact that the rise in pollutants amount and toxicity into our atmosphere is due to increase in anthropogenic sources at an astronomical rate. Here are some sources of pollutants those have lion's share in damaging Indian atmosphere.

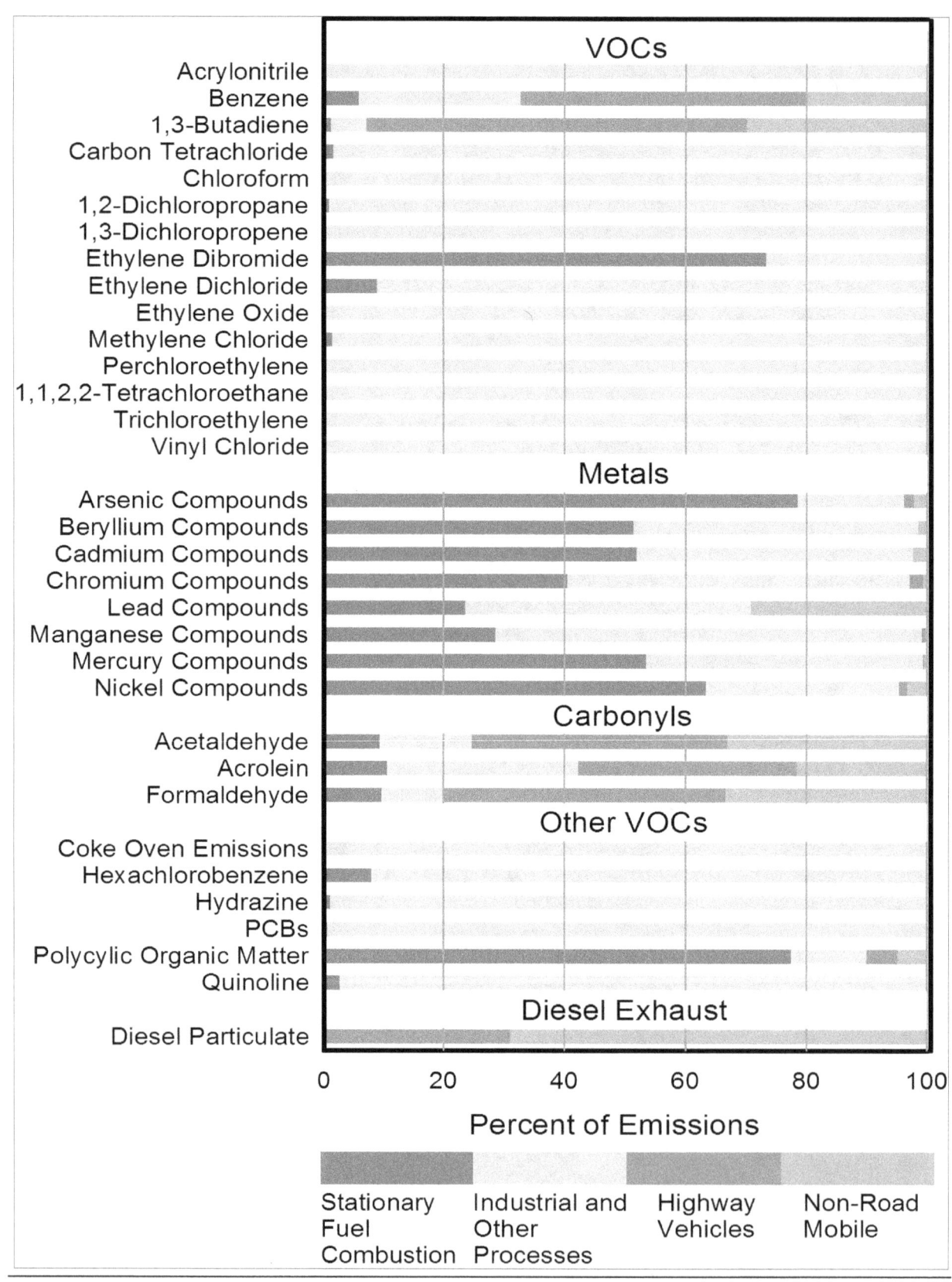

Fig 2.3: Pollutants and their sources. (www.epa.gov).

Burning plastic is not removing but adding to the pollution.

We often find burning plastics near streets and backyards .Most people who burn their plastic domestic waste do not realize how harmful this practice is to their health and to the environment. Current research indicates that backyard-burning of waste is far more harmful to our health than previously thought. It can increase the risk of heart disease, aggravate respiratory ailments such as asthma and emphysema, and cause rashes, nausea, or headaches, damages in the nervous system, kidney or liver, in the reproductive and development system. The burning of polystyrene polymers -such as foam cups, meat trays, egg containers, yogurt and other containers - releases styrene. Styrene gas can readily be absorbed through the skin and lungs. Not only these people who are burning the trash are exposed to these pollutants, but also their neighbours, children and families. The most dangerous emissions can be caused by burning plastics containing organochlor- based substances like PVC. When such plastics are burned, harmful quantities of dioxins, a group of highly toxic chemicals are emitted. Dioxins are the most toxic to the human organisms. They are carcinogenic and a hormone disruptor and persistent, and they accumulate in our body-fat and thus mothers give it directly to their babies via the placenta. Dioxins also settle on crops and in our waterways where they eventually wind up in our food, accumulate in our bodies and are passed on to our children. Surveys show that home burning of waste is widespread across rural areas all-over the world. Waste is either burned outside in the yard or garden, or inside in ovens. Waste that is burned can include paper, cardboard, food scraps and plastics, — essentially any materials that would otherwise be recycled or picked up by a waste collection company. Air emissions from home burning are released directly into the house or the atmosphere without being treated or filtered.

Pollutants released from burning plastic waste in a burn barrel are transported through the air either short or long distances, and are then deposited onto land or into bodies of water. A few of these pollutants such as mercury, polychlorinated biphenyls (PCBs), dioxins and furans persist for long periods of time in the environment and have a tendency to bio-accumulate which means they build up in predators at the top of the food web. Bioaccumulation of pollutants usually occurs indirectly through contaminated water and food rather than breathing the contaminated air directly. In wildlife, the range of effects associated with these pollutants includes cancer, deformed offspring, reproductive failure, immune diseases and subtle neurobehavioral effects. Humans can be exposed indirectly just like wildlife, especially through consumption of contaminated fish, meat and diary products.

(Fact sheet produced by:**WECF**, Women in Europe for a Common Future, Blumenstrasse 28,D – 80311 Munich, Germany,wecf@wecf.org,www.wecf.org.)

2.1 FUEL WOOD AND BIOMASS BURNING -COOKING FOOD, HAVING TOXICS.

Fig 2.1.1: Indian Chullah . (www.teriin.org)

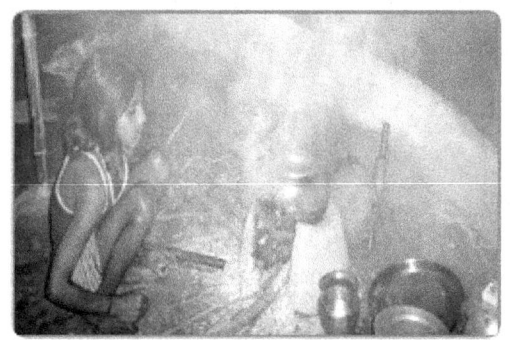

Pollution from different types of cooking stoves using , fuelwood, and other biomass fuels contributes to some extent, to the overall pollution load (UNEP "India; State of the Environment 2001") .Cooking fuel in rural India is prepared from a wet mix of dried grass, fuelwood pieces, hay, leaves and mostly cow/livestock dung. This mix is patted down into disc-shaped cakes, dried, and then used as fuel in stoves. When it burns, it produces smoke and numerous indoor air pollutants at concentrations 5 times higher than coal. Fuelwood and biomass burning is the primary reason for near-permanent haze and smoke observed above rural and urban India, and in satellite pictures of the country. Fuelwood and biomass cakes are used for cooking and general heating needs.Surveys suggest over 100 million households in India use rural stoves (chullahs) using biomass cakes, fuelwood and trash as cooking fuel every day, 2-3 times a day. Most of Indian population lives in rural areas so majority of Indians still use traditional fuels such as dried cow dung, agricultural wastes, and firewood as cooking fuel (Atmanand et al.2009).Clean burning fuels and electricity are unavailable in rural parts and small towns of India because of limited energy generation infrastructure. Moreover, as it is cheap to use chullah poor people prefer to use it for cooking rather expensive LPG.

This form of fuel is inefficient source of energy, its burning releases high levels of smoke, PM10 particulate matter, NOX, SOX, PAHs, polyaromatics, formaldehyde, carbon monoxide and other air pollutants. Scientific studies conclude biomass combustion in India is the country's dominant source of carbonaceous aerosols, emitting 0.25 teragram per year of black carbon into air, 0.94 teragram per year of organic matter, and 2.04 teragram per year of small particulates with diameter less than 2.5 microns. Biomass burning, as domestic fuel in India, accounts for about 3 times as much black carbon air pollution as all other sources combined, including vehicles and industrial sources (Reddy and Venkataraman .2002). Some reports, including one by the World Health Organization, claim 300,000 to 400,000 people die of indoor air pollution and carbon monoxide poisoning in India because of biomass burning and use of chullahs (The Times of India. December 3, 2009). India is the world's largest consumer of fuelwood, agricultural waste and biomass for energy purposes. Furthermore, India used 148.7 million tonnes coal replacement worth of fuelwood and biomass annually for domestic energy use. India's national average annual per capita consumption of fuel wood, agri waste and biomass cakes was 206 kilogram coal equivalent (Devendra Pandey .2002).

In 2010 terms, with India's population increased to about 1.2 billion, the country burns over 200 million tonnes of coal replacement worth of fuel wood and biomass every year to meet its energy need for cooking and other domestic use. The study found that the households consumed around 95 million tonnes of fuelwood, one-third of which was logs and the rest was twigs. Twigs were mostly consumed in the villages, and logs were more popular in cities of India. (Devendra Pandey .2002). The overall contribution of fuelwood, including sawdust and wood waste, was about 46% of the total, the rest being agri waste and biomass dung cakes. Traditional fuel (fuelwood, crop residue and dung cake) dominates domestic energy use in rural India and accounts for about 90% of the total. In urban areas, this traditional fuel constitutes about 24% of the total(Devendra Pandey ,2002).

Fuel wood, agri waste and biomass cake burning releases over 165 million tonnes of combustion products into India's indoor and outdoor air every year. To place this volume of emission in context, the Environmental Protection Agency (EPA) of the United States estimates that fire wood smoke contributes over 420,000 tonnes of fine particles throughout the United States – mostly during the winter months. United States consumes about one-tenth of fuelwood consumed by India, and mostly for fireplace and home heating purposes. The fuelwood quality in India is different than the dry firewood of the United States, and the Indian stoves in use are less efficient thereby producing more smoke and air pollutants per kilogram equivalent. India has less land area and less emission air space than the United States. In summary, the impact on indoor and outdoor air pollution by fuelwood and biomass cake burning is far worse in India.

Table 2.1.1: Common pollutants and their sources.(http://www.rrcap.ait.asia/male/manual/national/02chapter2.pdf)

Pollutants	Sources
Suspended particulate Matter SPMa	Automobile,power plants,boilers,Industries.
Chlorine	Chloro-alkali plants.
Fluoride	Fertilizer,aluminium refining
Sulphur dioxide	Power plants,boilers,sulphuric acid manufacture,ore refining,petroleum refining.
Lead	Ore refining,battery manufacturing,automobiles.
Oxides of nitrogen a	Automobiles,power plants,nitric NO,NO2 (NOx)acid manufacture,also a secondary pollutant.
Peroxyacetyl nitrate, PAN	Secondary pollutant
Formaldehyde	Secondary pollutant
Carbon monoxide$_a$	Automobiles
Hydrogen sulphide	Pulp and paper, petroleum refining.
Hydrocarbons	Automobiles,petroleum refining

a-Criteria pollutants source

2.2 VEHICULAR EMISSION –LIFE SUPPRESSION

Scientists and researchers unanimously agree on the fact that vehicles are major source of widespread air pollution in urban areas along with industries and thermal power plants. Vehicular emissions are of particular concern since these are ground level sources and thus have the maximum impact on the general population. People from different walks of life face it. As number of vehicles increasing every year in our country it manifest our dependence on them and more and more risk of accumulation of toxic pollutants in our ecological system The slow growth of road infrastructure and high growth of transport performance and number of vehicles all imply that Indian roads are reaching a saturation point in utilising the existing capacities.Apart from the concentration of vehicles in urban areas, other reasons for increasing vehicular pollution are the types of engines used, age of vehicles, congested traffic, poor road conditions, and outdated automotive technologies and traffic management systems (UNEP "India; State of the Environment 2001"). Beside petrol and LPG based small motor vehicles all heavy duty vehicles use diesel engine. It has been noted that popularity of diesel engine is rising in India. There is significant hike in number of diesel variant of vehicles on road every year. It is worrisome when several agencies including WHO have classified diesel exhaust as a carcinogen.

Diesel Emmision

In 1892, Rudolf Diesel invented the diesel "compression ignition" engine (http://www.nrdc.org/air/transportation/ebd/ebdinx.asp). A diesel engine operates by introducing air and fuel into the cylinder and compressing it to a point where the temperature is high enough to ignite the fuel without the necessity of a spark plug. This type of compression ignition system produces a significant amount of power and is fuel-efficient and durable. Diesel engines, however, emit more particulate matter per mile driven compared with gasoline engines of a similar weight class. The use of diesel engines spread throughout the United States and Europe after 1900, ultimately replacing steam-powered engines. Diesel engines operate on fairly inexpensive fuel oils and can withstand heavy loads at relatively low speeds. Conventional gasoline engines were unable to perform as well under heavy load conditions and required more expensive fuel. Due to the heavy weight of the early engines, diesel was used almost exclusively for heavy-duty power generation in marine transportation and to a limited extent in industrial establishments. The market for diesels broadened due to technological advances in the late 1930s that raised the operating speeds and decreased the engine weight, allowing the use of diesel engines for on-road applications. General Motors developed a two-cycle diesel engine that was suitable for railroad use, and was later adapted to drive trucks and buses. This was the beginning of a dependence on diesel for movement of freight and passengers, which has lasted through this century. The diesel engine is used mainly in trucks, buses, agricultural and other offroad equipment, locomotives, and ships.

Table 2.2. : Recommended emission factors by CPCB for on highway vehicles Vehicle type/ Control Technology.(CPCB.2010)

	CH4 (g/mi)	CH4 (g/km)
Gasoline Passenger Cars:		
Low Emission Vehicles	0.013	0.008
EPA Tier 1 a	0.02	0.012
EPA Tier 0 a	0.066	0.041
Oxidation Catalysts	0.133	0.083
Non-Catalyst	0.162	0.101
Uncontrolled	0.171	0.106
Diesel Passenger Cars:		
Advanced	0.001	0.001
Moderate	0.001	0.001
Uncontrolled	0.001	0.001
Diesel Light Duty Trucks:		
Advanced	0.001	0.001
Moderate	0.001	0.001
Uncontrolled	0.002	0.001
Diesel Heavy Duty Vehicles:		

Advanced	0.004	0.002
Moderate	0.004	0.002
Uncontrolled	0.004	0.002
Motorcycles:		
Non Catalysts Control	0.067	0.042
Uncontrolled	0.09	0.056

(The categories ―EPA Tier 0‖ and ―EPA Tier 1‖ were substituted for the early three-way catalyst and advanced three-way catalyst categories, respectively, as defined in the Revised 1996 IPCC Guidelines)
Advanced: EGR and modern electronic control of the fuel injection system are designated as advanced control technologies.
Moderate: Improved injection timing technology and combustion system design for light- and heavy-duty diesel vehicles (generally in place in model years 1983 to 1995) are considered moderate control technologies.
Uncontrolled: Not controlled over the combustion properties.
Note: Light and Heavy duty vehicles were tested on Heavy duty Chassis Dynamometer.
(Source: Anonymous, 2004, Update of Methane and Nitrous Oxide Emission Factors for On Highway Vehicles, Assessment and Standards Division Office of Transportation and -Air Quality, U.S. Environmental Protection Agency.)

Physico-Chemical Properties

The gaseous component of diesel exhaust is similar to the combustion products of other fuels. Although the adverse effects of diesel emissions are due both to the gaseous and particulate components, the toxicity of diesel exhaust is often expressed in relation to its particulate component. Complete and incomplete combustion of fuel in diesel engines results in a complex mixture of gases and particles composed of hundreds of organic and inorganic compounds. The physical and chemical characteristics of diesel exhaust are dependent on many factors such as the composition of the fuel, the characteristics of the engine and the conditions under which the diesel is burned. This section provides an overview of the different components of diesel exhaust. Table 2.2.2 lists the major constituents of diesel exhaust under standard condition.There are several toxic gaseous components in diesel exhaust such as formaldehyde, acetaldehyde ,acrolein benzene, 1,3-butadiene, carbon monoxide, polyaromatic hydrocarbons (PAHs),nitro-PAHs and dioxin compounds (http://www. toronto.ca/health /pdf/de_technical_ appendix.pdf).

Table 2.2.2. Percent Composition (by weight) of light-duty diesel engine exhaust (IPCS, 1996) under standard conditions.

Pollutant	Percent Composition
Carbon dioxide	7.1
Water vapour	2.6
Oxygen	15.0
Nitrogen	75.2
Carbon monoxide	0.03
Hydrocarbons	0.0007
Nitrogen oxides	0.03
Hydrogen	0.002
Sulphur dioxide	0.01
Sulphates	0.00016
Aldehydes	0.0014
Ammonia	0.00005
Particles	0.006

Diesel particulate matter (DPM) is the particle-phase of substances emitted in diesel exhaust. It refers to both the primary emissions and the secondary particles that are formed by atmospheric processes. Primary diesel particles are considered fresh after being emitted and undergo ageing (oxidation, nitration, or other chemical and physical changes) in the atmosphere.

Diesel exhaust particles are aggregates of primary spherical particles that consist of solid carbonaceous material and ash with associated adsorbed material. The particle portion of diesel exhaust contains elemental carbon (EC), organic carbon (OC), and small amounts of sulphate, nitrate, metals, trace elements, water, and other unidentified compounds. Elemental carbon usually makes up 50%-75% of the particles. Organic carbon makes up 19%-43% of the exhaust. It is composed of unburned fuel, engine oil, and small amounts of partial combustion and pyrolysis products. Polyaromatic hydrocarbons make up less than 1% of diesel exhaust particle mass.

Carbonaceous matter refers to all carbon-containing compounds in diesel particles, and includes the elemental and organic carbon. Organic carbon is made up of compounds containing carbon and hydrogen. The soluble organic fraction (SOF) is the portion of diesel particulate matter that can be extracted into solution. About one quarter of SOF is unburned fuel and three quarters is unburned engine lubrication oil. Partial combustion and pyrolysis products represent a very small fraction of the mass of SOF. Soot is the insoluble portion of diesel particle matter formed by clusters of elemental carbon and organic carbon particles.

A large number of elements and metals have been detected in diesel exhaust (http://www.toronto.ca/health/pdf/de_technical_appendix.pdf). They include barium, calcium, chlorine, chromium, copper, iron, lead, manganese, mercury, nickel, phosphorus, sodium, silicon, and zinc. These make up less than 1% of particle mass. Most of the sulphur in the fuel is oxidized to sulphur dioxide (SO_2), but about 1-4% is oxidized and then converted to sulphate and sulphuric acid in the exhaust. The amount of SO_2 emitted is related to the sulphur content of the fuel. Non-road equipment uses fuel containing more sulphur than on-road diesel engines. The maximum allowable sulphur content in diesel is being reduced. Vehicles tested using low-sulphur fuel were found to have a sulphate content of only about 1%. Water content is about 1.3 times the amount of sulphate. About 1-20% of total particle mass in diesel exhaust is in the ultra-fine size range ($PM_{2.5}$). The majority of these ultra-fine particles have an average size of 0.02 microns (range of 0.005-0.05 microns). They account for 50%-90% of the total number of particles. These very small particles are largely composed of sulphate and/or sulphate with condensed organic carbon.

Approximately 80%-95% of the mass of particles in diesel exhaust is made up of fine particles (PM_{10}) with an average diameter of about 0.2 microns size range (range from 0.05-1.0 microns). The particles in this range are composed of spherical elemental carbon cores on which are adsorbed organic compounds, sulphate, nitrate and trace elements. Their large surface area makes them excellent carriers for the adsorbed compounds, which can effectively reach the lowest parts of the lung.

PAH and nitro-PAH make up about 1% of the particulate component of diesel exhaust. Differences in engine type and make, general engine condition, fuel composition and test conditions can influence the emissions levels of PAH. Increasing the aromatic content of the fuel may also increase PAH emissions (http://www.toronto.ca/health/pdf/de_technical_appendix.pdf).

Table 2.2.3: Substances in Diesel Exhaust Listed by Cal EPA as Toxic Air Contaminants.
(http://www.nrdc.org/air/transportation/ebd/ebdinx.asp)

acetaldehyde	inorganic lead
acrolein	manganese compounds
aniline	mercury compounds
antimony compounds	methanol
arsenic	methyl ethyl ketone
benzene	naphthalene
beryllium compounds	nickel
biphenyl	4-nitrobiphenyl
bis[2-ethylhexyl]phthalate	phenol
1,3-butadiene	phosphorus
cadmium	polycyclic organic matter, including
chlorine	polycyclic aromatic hydrocarbons (PAHs)
chlorobenzene	and their derivatives
chromium compounds	propionaldehyde
cobalt compounds	selenium compounds
creosol isomers	styrene
cyanide compounds	toluene
dibutylphthalate	xylene isomers and mixtures
dioxins and dibenzofurans	o-xylenes
ethyl benzene	m-xylenes
formaldehyde	p-xylenes

Note: California Health and Safety Code section 39655 defines a "toxic air contaminant" as "an air pollutant which may cause or contribute to an increase in mortality or in serious illness, or which may pose a present or potential hazard to human health."

2.3 COAL –COLORING IN ITS COLOR.

Fig 2.3.1.(a)Increase in consumption of Coal and Natural Gas all around the world. (© 2013 Worldwatch Institut).(b)Burning coal (ec.europa.eu).

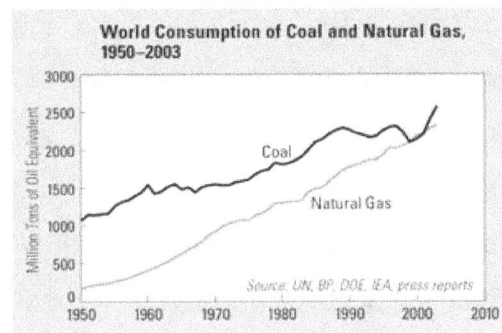

(a)   (b)

The rapid growth in coal use in China and India, where pollution controls are minimal, is adding to local and long-distance pollution (© 2013 Worldwatch Institute).Moreover, India's coal has a very high in ash content (24%–45%). The increased dependence of the power sector on an inferior quality coal has been associated with emissions from power plants in the form of particulate matter, toxic elements, fly ash, oxides of nitrogen, sulphur and carbon besides ash, which required vast stretches of land for disposal (UNEP "India; State of the Environment 2001").

2.4 INDUSTRIAL EXHAUST: EXHAUSTING LIVES

Fig.2.4.1.Industrial exhaust.(www.deepgreenresistance.org)

Industrial pollution has adversely affected our environment in many ways for the last two centuries and continues to increase globally. It is responsible for health problems,climatic problems,and loss of biodiversity.It leads to toxification of environmental sites, whereby the organisms living in the ecosystem are damaged because of the poisonous nature of many pollutants(http://web.mit.edu/12.000/www/m2015/2015/solutions_for_industrial_pollution.htm).Air pollution in developing countries is derived not only from stack emission of pollutants from relatively large industries, like iron and steel, non-ferrous metals and petroleum products industries, but also from fugitive emission of pollutants from small-scale factories, such as cement mills, lead refineries, chemical fertilizer and pesticide factories and so on, where inadequate pollution control measures exist and pollutants are allowed to escape to the atmosphere. Since industrial activities always involve energy generation, the combustion of fossil fuels is a main source of air pollution in the developing countries, where coal is widely used not only for industrial, but also for domestic consumption. The kinds of air pollutants emitted vary from industry to industry. The concentrations of different pollutants in the atmosphere also vary widely from process to process, and from place to place with different geographic and climatic conditions. It is difficult to estimate specific exposure levels of various pollutants from different industries to the general population in developing countries, as elsewhere. It effects both general population and workers as per exposure quantity and duration.

The exposure levels of the general population in developing countries are usually higher than that in developed countries, where air pollution is more strictly controlled and resident areas are usually far from industries.A large number of epidemiological studies have already showed the close association of reduction in pulmonary function and increased incidence of chronic respiratory diseases among residents with long-term exposure to the common air pollutants.

A case study of air pollution effects on the health of 480 primary school children in Cubatao, Brazil, where large quantities of mixed pollutants were emitted from 23 industries (steel mill, chemical industries, cement factory, fertilizer plants, etc.), showed that 55.3% of the children had decreases in pulmonary function. Another example of health effects of air pollution appeared in the Ulsan/Onsan special industrial zone,

Republic of Korea, where many large-scale plants (mainly petrochemical plants and metal refineries) are concentrated. Local residents complained of a variety of health problems, particularly of the nervous system disorder called "Onsan Disease"(http://www.ilo.org/oshenc/part-vii/environmental-health-hazards/item/497-industrial-pollution-in-developing-countries).

India is a newly industrialized country, wherein the transition from agriculture-based growth to industry-based development happened only post-independence (after 1947), and since then, industrialization has played a major role in the country's modernization and socio-economic development. In such a short period India has got a status of being one of the ten most industrialised nations of the world (UNEP "India; State of the Environment 2001"). Though agriculture and allied sectors still account for 16.1% of the GDP, the share of industries has gone up to 28.6% while services sector accounts for 55.3% share in GDP (http://marketspace.thinktosustain.com/2011/10/most-polluted-industrial-clusters-of-india-a-review/#.UfDEJ2Tn_Dc). Industries are now a backbone to the country's economy, bringing about rapid socio-economic development and significant lifestyle changes for its people. But when industrialization took root in India, little did anyone know that the industries, which were proving to be a boon to quite a large number of people, would also spread their deadly tentacles and engulf the environment, eventually becoming a major cause of air, water and land pollution. It has brought with it unwanted and unanticipated consequences such as unplanned urbanisation, pollution and the risk of accidents (UNEP "India; State of the Environment 2001").

Industrial pollution is pollution which can be directly linked with industry, in contrast to other pollution sources. This form of pollution is one of the leading causes of pollution worldwide; in the United States, for example, the Environmental Protective Agency (US EPA) estimates that up to 50% of the nation's pollution is caused by industry.

Industrial survey statistics tell that more than one-third (36.3 per cent) of the total value added by the raw materials through manufacture in the factory sector of the country comes from Maharashtra (23.66 per cent) and Gujarat (12.64 per cent), easily making them the two most industrialized states of India. Governments of both the states claim they have created immense prosperity in the region. But statistics do not tell the real story of thousands of workers, farmers and general population, who suffer at the hands of pollution created by the industries while creating this prosperity.India's Central Pollution Control Board (CPCB), in conjunction with IIT-Delhi, surveyed 88 industrial clusters across the country, and found 43 "critically polluted" (with scores above 70 on a 100-point scale) while 32 were "severely polluted" (scores 60-70). The CPCB (Central Pollution Control Board) has identified seventeen categories of industries (large and medium scale) as significantly polluting and the list includes highly air polluting industries such as integrated iron and steel, thermal power plants, copper/zinc/

aluminium smelters, cement, oil refineries, petrochemicals, pesticides and fertiliser units.

Thermal power plants are major sources of SPM, SO2 and NOx. Depending upon the type of fuel used, emission of one or more of these pollutants may be of environmental significance.A large amount of SPM as fly ash is emitted from coal fired plants, particularly if the ashcontent of coal is high and a fly ash removal unit, such as, an electrostatic precipitation(ESP) is not used.Oxides of nitrogen formed in combustion processes are usually due to either thermal fixation of atmospheric nitrogen in combustion air or to the conversion of chemically bound nitrogen in the fuel. Thermal fixation occurs when combustion temperature is above 1600°C. For natural gas and distillate oil nearly all NO results from thermal fixation. For residue oil and coal, the contribution to NO emission from fuel bound nitrogen may be significant.The concentration of NOx formed increases with increase in excess oxygen maintained in the combustion process and with the increase in temperature of the furnace. For coal based thermal power plants in India, it ranges between 100 and 200 mg/Nm3 in the flue gas. In the case of natural gas and liquid fuels, the emission limits for flue gases prescribed in European countries is in the range of 200 - 400 mg/Nm3

In cement manufacturing , raw materials include lime, silica, aluminum, iron sand, clay, shale,and blast furnace slag follow the process of mining, crushing, grinding, and calcining in a long cylindrically shaped oven or kiln. Air pollutants can originate at several operations as during raw material crushing, grinding particulates, kiln operation and cooling particulates, CO, SO2 ,NOx, HC and product grinding and packaging particulates.

The paper mills play very important role in economic and industrial growth of India (http://www.clarion.ind.in/index.php/clarion/article/view/37/53). There are about 215 Paper mills in India comprising 179 small paper mills with installing capacity of 7,12,340 tons per anum (TPA) and 36 large paper mills with installing capacity of 13,29,160 TPA. The total installed capacity is 2.4 million TPA, to cater the domestic requirement of the country. The small paper mills contribute about 35% and large paper mills contribute about 65% of the total production of paper in the country. Most of the paper mills, particularly large Paper mills are dependent on forest based raw materials like bamboo and forest wood for manufacture of paper.

Paper Mills have direct influence on ecology and biodiversity due to large scale consumption of forest products and simultaneous release of huge amount of toxic liquid, solid and gaseous waste in to the environment. The following shows the volume of consumption of natural raw materials and chemicals to produce one ton of writing paper by a Paper mill (http://www.clarion.ind.in/index.php/clarion/article/view/37/53).

1. *Bamboo (moisture free) – 2.2 metric ton*
2. *Caustic soda/salt cake – 150 Kg*
3. *Chlorine – 120 Kg*
4. *Talcum powder – 200 Kg*
5. *Alum – 50 Kg*
6. *Rosin -10 Kg*
7. *Dyes – 50 gm*
8. *Power – 1800 KW*

9. Coal – 2 tons.
10. Furnace oil – 40 liters
11. Water – 250 cubic meter (Baruah et al. 1996 a)

A very well known case is of Baikal pulp and paper mill(BPPM),Russia. Environmentalists had long campaigned against the industrial pollution of wildlife around Baikal and the lake itself, which features on the UNESCO list of world heritage sites. According to Greenpeace, the BPPM is an outrageously obsolete industrial facility that pollutes Lake Baikal with chlorine and sulfur compounds and also dioxin - a highly toxic chemical. All affect the food chain. The amount of waste dumped annually is estimated at 27 million tonnes.Considering all these facts Prime Minister Dmitry Medvedev visited the Irkutsk Region in June 2013 and declared that the mill would be out of operation in two years' time and its personnel given jobs in the local tourist industry.(http://indrus.in/society/2013/06/20/polluting_baikal_pulp_and_paper_mill_to_be_shut_26271.html.)

Fig 2.4. Evil-looking froth in the Aril River in Uttar Pradesh.

Similarly, in Amarpurkashi (a village in Moradabad district in the state of Uttar Pradesh, Northern India),in the last ten years, the local paper mill has caused horrific pollution. At first, it was just the Aril River which became sluggish, black, acidic and covered in an evil-looking froth. Then it spread to farmers' adjacent fields, ruining their crops. Next, the mill owners dumped live ash on the roadside where unsuspecting cyclists and pedestrians walked or rode through what they assumed were cold ashes. Many villagers suffered horrendous burns as a result. Now, the stink of chemical effluents pollutes the air of the surrounding villages while black dust from the factory chimney blows far and wide, settling on everything. The water table has dropped dramatically as the factory uses huge amounts of water and all the roadside ponds have dried up. The underground water supply has also become polluted causing a sudden rise in the number of people suffering from jaundice. Those farmers who lose their crops every year have never been compensated while villagers are forced to pay for ever-deeper borings to ensure a clean water supply (http ://www.vri-online.org.uk/apk/pollution-campaign.php).

3.Pollutants

Table 3.1: Pollution Level Classification. (CPCB.2012)

Pollution level	Annual Mean Concentration Range (µg/m$_3$)					
	Industrial, Residential,Rural & others areas			Ecologically Sensitive Area		
	SO_2	NO_2	PM_{10}	SO_2	NO_2	PM_{10}
Low (L)	0-25	0-20	0-30	0-10	0-15	0-30
Moderate (M)	26-50	21-40	31-60	11-20	16-30	31-60
High (H)	51-75	41-60	61-90	21-30	31-45	61-90
Critical (C)	>75	>60	>90	>30	>45	>90

1. SUSPENDED PARTICULATE MATTER

Particulates, alternatively referred to as particulate matter (PM), atmospheric particulate matter, or fine particles, are tiny particles of solid or liquid suspended in a gas or aerosol to particles and the gas together. However, it is common to use the term aerosol to refer to the particulate component alone (Seinfeld, John; Spyros Pandis ,1998).

Fig3.1.1:This diagram shows the size distribution in micrometres of various types of atmospheric particulate matter. It also shows the different types of particulates in the atmosphere. (Wikipedia)

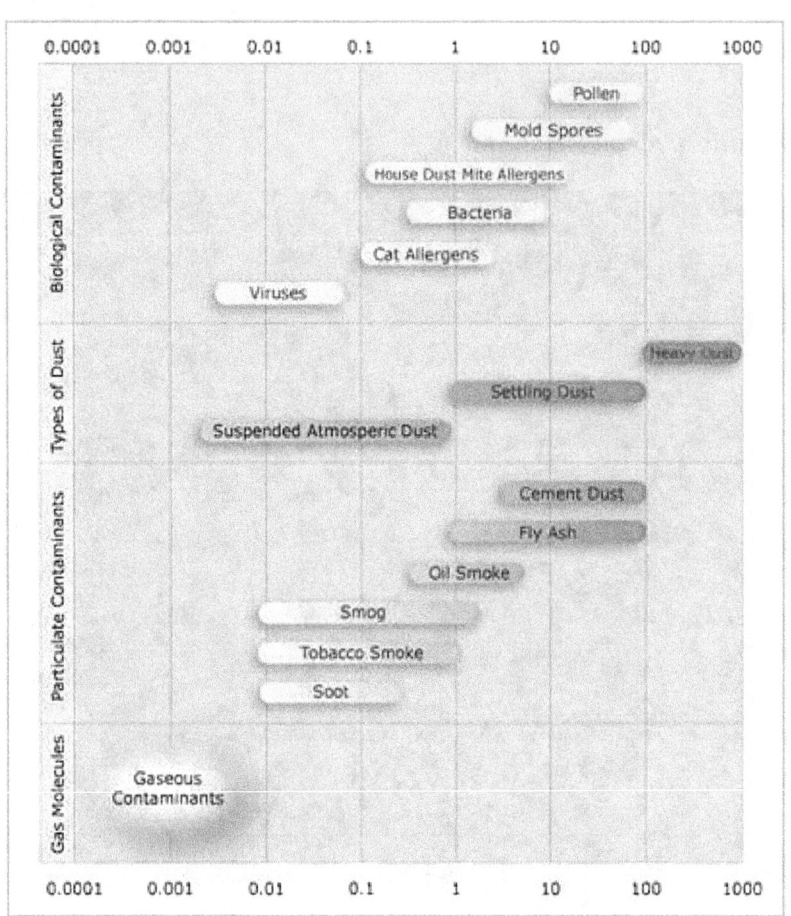

Sources of particulate matter can be manmade or natural. They can adversely affect human health and also have impacts on climate and precipitation. Some particulates occur naturally, originating from volcanoes, dust storms, forest and grassland fires, living vegetation, and sea spray. Human activities, such as the burning of fossil fuels in vehicles, power plants , agricultural activities, construction, demolition ,and various industrial processes also generate significant amounts of particulates. More than 98 percent of the total number of particles in diesel exhaust is PM2.5. Coal combustion in developing countries is the primary method for heating homes and supplying energy. Because salt spray over the oceans is the overwhelmingly most common form of particulate in the atmosphere, anthropogenic aerosols—those made by human activities—currently account for about 10 percent of the total mass of aerosols in our atmosphere (Mary Hardin and Ralph Kahn. "Aerosols and Climate Change").Under various conditions of their generation, they are also called by other names such as dust, fume, smoke , mist ,metallic oxides, and pollen. Dust usually refers to particles in the range of 1 to 200 μm size. Fume is very fine solid or liquid particles arising from chemical reactions or condensation of gases. Smoke refers to finely divided particles resulting from incomplete combustion of substances such as coal, petroleum, etc. Larger particles that enter the respiratory system are trapped by hairs and lining of nose or can be captured by mucus in upper respiratory tract and worked back to the throat by cilia and removed by spitting. Particles in the range of 0.1μm to 10μm are of, most interest from health viewpoint. According to results of studies conducted by NAMP (Nation Air Monitoring Program, India)in 2010 ,most of the Indian cities are critically polluted by PM$_{10}$. (National Ambient Air Quality Status & Trends In India-2010, CPCB January,2012)

Table 3.1.1: Ten states with highest PM10 values (annual average) during 2010 (residential / industrial / rural / other & ecologically sensitive area). (CPCB.2012)

Sl. No.	State	Min	Max	Annual average (μg/m$_3$)
1	Delhi	46	748	261*
2	Jharkhand	84	398	193*
3	Punjab	115	299	187*
4	Uttar Pradesh	96	484	181*
5	Bihar	92	504	171*
6	Chattisgarh	92	263	169*
7	Rajasthan	32	576	168*
8	Haryana	185	149	137*
9	Uttrakhand	36	656	118*
10	Madhya Pradesh	24	308	110*

> **Finding of National Air Quality Program (NAMP) conducted all across the country in 2010 says:**
>
> - Cities like Badlapur and Ulhasnagar (Maharashtra), Asansol, Durgapur, Barrackpur, Howrah, Kolkata Raniganj and Sankrail (West Bengal) are critical with respect to both NO_2 and PM_{10}
>
> - State capital cities like Patna, Raipur, Delhi, Ahmedabad, Ranchi, Bhopal, Mumbai, Amritsar, Jaipur, Lucknow, and Kolkata are critical with respect to PM_{10}
>
> - Industrial cities like Bhilai, Korba, Ahmedabad, Faridabad, Jamshedpur, Jharia, Sindri, Ludhiana, Muradabad, Rourkela, Indore, Kota, Kanpur, Asansol, Durgapur, Howrah are critical with respect to PM_{10}

Particulate Matter (PM or "aerosols") is usually categorized into these groups based on the aerodynamic diameter of the particles(http://www.aiha.org/abs05/po105.htm, http://www.greenfacts.org/glossary/pqrs/PM10-PM2.5-PM0.1.htm):

1. Particles less than 100 microns, which are also called "inhalable" since they can easily enter the nose and mouth.
2. Particles less than 10 microns (PM10, often labeled "fine" in Europe). These particles are also called "thoracic" since they can penetrate deep in the respiratory system.
3. Particles less than 4 microns. These particles are often called "respirable" (http://www.aiha.org/abs05/po105.htm) because they are small enough to pass completely through the respiratory system and enter the bloodstream.
4. Particles less than 2.5 microns (PM2.5, labeled "fine" in the US).

5. Particles less than 0.1 microns (PM0.1, "ultrafine") (Daly, A. and P. Zannetti. 2007).

Fig 3.1.2: Particulate matter.

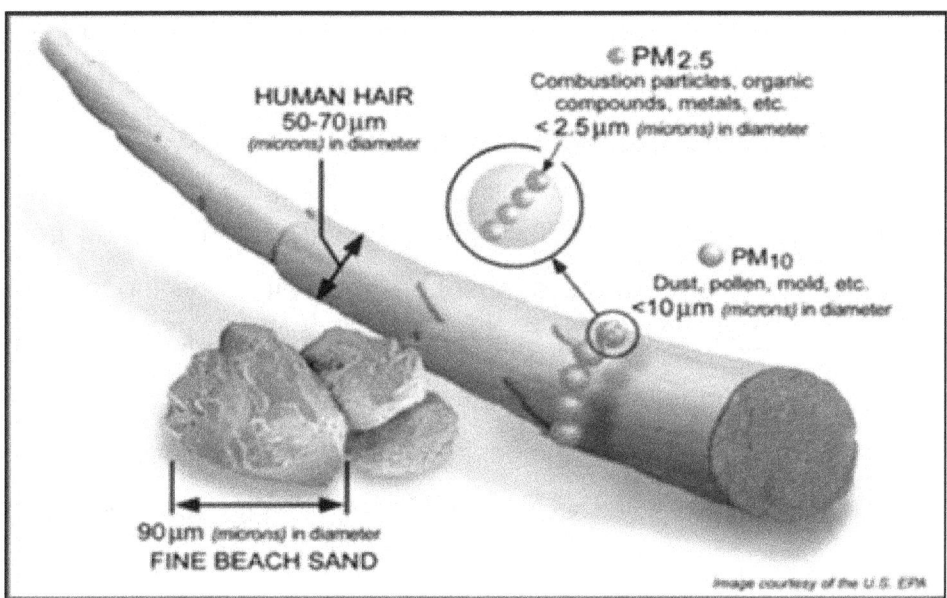

Aerosol Composition

The composition of aerosols and particles depends on their source. Wind-blown mineral dust (UNEP. 2001) tends to be made of mineral oxides and other material blown from the Earth's crust; this particulate is light-absorbing (National Academy of Sciences of the United States of America January 30, 2012),. Sea salt(UNEP. 2001) is considered the second-largest contributor in the global aerosol budget, and consists mainly of sodium chloride originated from sea spray; other constituents of atmospheric sea salt reflect the composition of sea water, and thus include magnesium, sulfate, calcium, potassium, etc. In addition, sea spray aerosols may contain organic compounds, which influence their chemistry.

Secondary particles derive from the oxidation of primary gases such as sulfur and nitrogen oxides into sulfuric acid (liquid) and nitric acid (gaseous). The precursors for these aerosols—i.e. the gases from which they originate—may have an anthropogenic origin (from fossil fuel or coal combustion) and a natural biogenic origin. In the presence of ammonia, secondary aerosols often take the form of ammonium salts; i.e. ammonium sulfate and ammonium nitrate (both can be dry or in aqueous solution); in the absence of ammonia, secondary compounds take an acidic form as sulfuric acid (liquid aerosol droplets) and nitric acid (atmospheric gas). Secondary sulfate and nitrate aerosols are strong light-scatterers(UNEP. 2001). This is mainly because the presence of sulfate and nitrate causes the aerosols to increase to a size that scatters light effectively.

Organic matter (OM) can be either primary or secondary, the latter part deriving from the oxidation of VOCs; organic material in the atmosphere may either be biogenic or anthropogenic. Organic matter influences the atmospheric radiation field by both scattering and absorption. Another important aerosol type is elemental carbon (EC, also known as black carbon, BC): this aerosol type includes strongly light-absorbing material and is thought to yield large positive radiative forcing. Organic matter and elemental carbon together constitute the carbonaceous fraction of aerosols (UNEP. 2001). Secondary organic aerosols, tiny "tar balls" resulting from combustion products of internal combustion engines, have been identified as a danger to health (Felicity Barringer ,February 18, 2012).

The chemical composition of the aerosol directly affects how it interacts with solar radiation. The chemical constituents within the aerosol change the overall refractive index. The refractive index will determine how much light is scattered and absorbed.

The composition of particulate matter that generally causes visual effects such as smog consists of sulphur dioxide, nitrogen oxides, carbon monoxide, mineral dust, organic matter, and elemental carbon also known as black carbon or soot. The particles are hydroscopic due to the presence of sulphur, and SO_2 is converted to sulphate when high humidity and low temperatures are present. This causes the reduced visibility and yellow color (The World Bank. 2010).

Table 3.1.2: The original list of hazardous air pollutants as follows. (EPA)

CAS Number	Chemical Name
75070	Acetaldehyde
60355	Acetamide
75058	Acetonitrile
98862	Acetophenone
53963	2-Acetylaminofluorene
107028	Acrolein
79061	Acrylamide
79107	Acrylic acid
107131	Acrylonitrile
107051	Allyl chloride
92671	4-Aminobiphenyl
62533	Aniline
90040	o-Anisidine
1332214	Asbestos
71432	Benzene (including benzene from gasoline)
92875	Benzidine
98077	Benzotrichloride
100447	Benzyl chloride
92524	Biphenyl
117817	Bis(2-ethylhexyl)phthalate (DEHP)
542881	Bis(chloromethyl)ether
75252	Bromoform
106990	1,3-Butadiene
156627	Calcium cyanamide
105602	Caprolactam
133062	Captan
63252	Carbaryl
75150	Carbon disulfide
56235	Carbon tetrachloride
463581	Carbonyl sulfide
120809	Catechol
133904	Chloramben
57749	Chlordane
7782505	Chlorine
79118	Chloroacetic acid
532274	2-Chloroacetophenone
108907	Chlorobenzene
510156	Chlorobenzilate
67663	Chloroform
107302	Chloromethyl methyl ether
126998	Chloroprene
1319773	Cresols/Cresylic acid (isomers and mixture)
95487	o-Cresol
108394	m-Cresol
106445	p-Cresol
98828	Cumene
94757	2,4-D, salts and esters
3547044	DDE
334883	Diazomethane
132649	Dibenzofurans
96128	1,2-Dibromo-3-chloropropane
84742	Dibutylphthalate
106467	1,4-Dichlorobenzene(p)
91941	3,3-Dichlorobenzidene
111444	Dichloroethyl ether (Bis(2-chloroethyl)ether)
542756	1,3-Dichloropropene
62737	Dichlorvos

CAS #	Chemical
111422	Diethanolamine
121697	N,N-Dimethylaniline
64675	Diethyl sulfate
119904	3,3-Dimethoxybenzidine
60117	Dimethyl aminoazobenzene
119937	3,3'-Dimethyl benzidine
79447	Dimethyl carbamoyl chloride
68122	Dimethyl formamide
57147	1,1-Dimethyl hydrazine
131113	Dimethyl phthalate
77781	Dimethyl sulfate
534521	4,6-Dinitro-o-cresol, and salts
51285	2,4-Dinitrophenol
121142	2,4-Dinitrotoluene
123911	1,4-Dioxane (1,4-Diethyleneoxide)
122667	1,2-Diphenylhydrazine
106898	Epichlorohydrin (l-Chloro-2,3-epoxypropane)
106887	1,2-Epoxybutane
140885	Ethyl acrylate
100414	Ethyl benzene
51796	Ethyl carbamate (Urethane)
75003	Ethyl chloride (Chloroethane)
106934	Ethylene dibromide (Dibromoethane)
107062	Ethylene dichloride (1,2-Dichloroethane)
107211	Ethylene glycol
151564	Ethylene imine (Aziridine)
75218	Ethylene oxide
96457	Ethylene thiourea
75343	Ethylidene dichloride (1,1-Dichloroethane)
50000	Formaldehyde
76448	Heptachlor
118741	Hexachlorobenzene
87683	Hexachlorobutadiene
77474	Hexachlorocyclopentadiene
67721	Hexachloroethane
822060	Hexamethylene-1,6-diisocyanate
680319	Hexamethylphosphoramide
110543	Hexane
302012	Hydrazine
7647010	Hydrochloric acid
7664393	Hydrogen fluoride (Hydrofluoric acid)
7783064	Hydrogen sulfide
123319	Hydroquinone
78591	Isophorone
58899	Lindane (all isomers)
108316	Maleic anhydride
67561	Methanol
72435	Methoxychlor
74839	Methyl bromide (Bromomethane)
74873	Methyl chloride (Chloromethane)
71556	Methyl chloroform (1,1,1-Trichloroethane)
78933	Methyl ethyl ketone (2-Butanone)
60344	Methyl hydrazine
74884	Methyl iodide (Iodomethane)
108101	Methyl isobutyl ketone (Hexone)
624839	Methyl isocyanate
80626	Methyl methacrylate
1634044	Methyl tert butyl ether
101144	4,4-Methylene bis(2-chloroaniline)
75092	Methylene chloride (Dichloromethane)
101688	Methylene diphenyl diisocyanate (MDI)
101779	4,4'-Methylenedianiline

91203	Naphthalene
98953	Nitrobenzene
92933	4-Nitrobiphenyl
100027	4-Nitrophenol
79469	2-Nitropropane
684935	N-Nitroso-N-methylurea
62759	N-Nitrosodimethylamine
59892	N-Nitrosomorpholine
56382	Parathion
82688	Pentachloronitrobenzene (Quintobenzene)
87865	Pentachlorophenol
108952	Phenol
106503	p-Phenylenediamine
75445	Phosgene
7803512	Phosphine
7723140	Phosphorus
85449	Phthalic anhydride
1336363	Polychlorinated biphenyls (Aroclors)
1120714	1,3-Propane sultone
57578	beta-Propiolactone
123386	Propionaldehyde
114261	Propoxur (Baygon)
78875	Propylene dichloride (1,2-Dichloropropane)
75569	Propylene oxide
75558	1,2-Propylenimine (2-Methyl aziridine)
91225	Quinoline
106514	Quinone
100425	Styrene
96093	Styrene oxide
1746016	2,3,7,8-Tetrachlorodibenzo-p-dioxin
79345	1,1,2,2-Tetrachloroethane
127184	Tetrachloroethylene (Perchloroethylene)
7550450	Titanium tetrachloride
108883	Toluene
95807	2,4-Toluene diamine
584849	2,4-Toluene diisocyanate
95534	o-Toluidine
8001352	Toxaphene (chlorinated camphene)
120821	1,2,4-Trichlorobenzene
79005	1,1,2-Trichloroethane
79016	Trichloroethylene
95954	2,4,5-Trichlorophenol
88062	2,4,6-Trichlorophenol
121448	Triethylamine
1582098	Trifluralin
540841	2,2,4-Trimethylpentane
108054	Vinyl acetate
593602	Vinyl bromide
75014	Vinyl chloride
75354	Vinylidene chloride (1,1-Dichloroethylene)
1330207	Xylenes (isomers and mixture)
95476	o-Xylenes
108383	m-Xylenes
106423	p-Xylenes
0	Antimony Compounds
0	Arsenic Compounds (inorganic including arsine)
0	Beryllium Compounds
0	Cadmium Compounds
0	Chromium Compounds
0	Cobalt Compounds
0	Coke Oven Emissions
0	Cyanide Compounds1

0	Glycol ethers2
0	Lead Compounds
0	Manganese Compounds
0	Mercury Compounds
0	Fine mineral fibers3
0	Nickel Compounds
0	Polycyclic Organic Matter4
0	Radionuclides (including radon)5
0	Selenium Compounds

NOTE: For all listings above which contain the word "compounds" and for glycol ethers, the following applies: Unless otherwise specified, these listings are defined as including any unique chemical substance that contains the named chemical (i.e., antimony, arsenic, etc.) as part of that chemical's infrastructure.

1 X'CN where X = H' or any other group where a formal dissociation may occur. For example KCN or Ca(CN)2

2 Includes mono- and di- ethers of ethylene glycol, diethylene glycol, and triethylene glycol R-(OCH2CH2)n -OR' where

n = 1, 2, or 3

R = alkyl or aryl groups

R' = R, H, or groups which, when removed, yield glycol ethers with the structure: R-(OCH2CH)n-OH. Polymers are excluded from the glycol category. (See Modification)

3 Includes mineral fiber emissions from facilities manufacturing or processing glass, rock, or slag fibers (or other mineral derived fibers) of average diameter 1 micrometer or less.

4 Includes organic compounds with more than one benzene ring, and which have a boiling point greater than or equal to 100 ° C.

5 A type of atom which spontaneously undergoes radioactive decay.

2. SULFUR OXIDES (SO$_X$).

Sulfur oxides especially sulphur dioxide, a chemical compound with the formula SO$_2$. This compound is colorless, but has a suffocating, pungent odor. SO$_2$ is produced by volcanoes and in various industrial processes. Further oxidation of SO$_2$, usually in the presence of a catalyst such as NO$_2$, forms H$_2$SO$_4$, and thus acid rain (Holleman, A. F, Wiberg, E,2001). Sulfur dioxide emissions are also a precursor to particulates in the atmosphere. Sulfur dioxide is a noticeable component in the atmosphere, especially following volcanic eruptions.

- Two resonance structures of sulfur dioxide.

The primary source of SO2 is the combustion of sulfur-containing fuels (e.g., oil and coal). Since coal and petroleum often contain sulfur compounds, their combustion generates sulfur dioxide.

Sulfur dioxide is the product of the burning of sulfur or of burning materials that contain sulfur:

$$S_8 + 8\ O_2 \rightarrow 8\ SO_2$$

The combustion of hydrogen sulfide and organosulfur compounds proceeds similarly.

$$2\ H_2S + 3\ O_2 \rightarrow 2\ H_2O + 2\ SO_2$$

The roasting of sulfide ores such as pyrite, sphalerite, and cinnabar (mercury sulfide) also releases SO_2 (Shriver, 2010)

$$4\ FeS_2 + 11\ O_2 \rightarrow 2\ Fe_2O_3 + 8\ SO_2$$
$$2\ ZnS + 3\ O_2 \rightarrow 2\ ZnO + 2\ SO_2$$
$$HgS + O_2 \rightarrow Hg + SO_2$$
$$4\ FeS + 7O_2 \rightarrow 2\ Fe_2O_3 + 4\ SO_2$$

A combination of these reactions is responsible for the largest source of sulfur dioxide, volcanic eruptions. These events can release millions of tonnes of SO_2.

Sulfur dioxide can also be a by-product in the manufacture of calcium silicate cement: $CaSO_4$ is heated with coke and sand in this process:

$$2\ CaSO_4 + 2\ SiO_2 + C \rightarrow 2\ CaSiO_3 + 2\ SO_2 + CO_2$$

Up until the 1970s, commercial quantities of sulfuric acid and cement were produced by this process in Whitehaven Britain. Upon being mixed with shale or marl, and roasted, the sulfate liberated sulfur dioxide gas, used in sulfuric acid production, the reaction also produced calcium silicate, a precursor in cement production.(http://www.lakestay.co.uk/whitehavenmininghisto ry.html). On a laboratory scale, the action of hot sulfuric acid on copper turnings produces sulfur dioxide.

$$Cu + 2\ H_2SO_4 \rightarrow CuSO_4 + SO_2 + 2\ H_2O$$

Sulfite results from the reaction of aqueous base and sulfur dioxide. The reverse reaction involves acidification of sodium metabisulfite:

$$H_2SO_4 + Na_2S_2O_5 \rightarrow 2\ SO_2 + Na_2SO_4 + H_2O$$

Sulfur dioxide is an intermediate in the production of sulfuric acid, being converted to sulfur trioxide, and then to oleum, which is made into sulfuric acid. Sulfur dioxide is sometimes used as a preservative , in winemaking, (The Food Standards Agency website)(MoreThanOrganic.com) as reductant,as a decolorizer .

Sulphur dioxide when released in the atmosphere can also convert to SO3.When SO3 is inhaled it is likely to be absorbed in moist passages of respiratory tract. When it is entrained in an aerosol, however, it may reach far deeper into the lungs This is one of the causes for concern over the environmental impact of the use of these fuels as power sources.

Sulfur trioxide (also spelled sulphur trioxide) is the chemical compound with the formula SO_3. In the gaseous form, this species is a significant pollutant, being the primary agent in acid rain (Lide, David R., 2006). It is prepared on massive scales as a precursor to sulfuric acid.

$$SO_3\ (g) + H_2O\ (l) \rightarrow H_2SO_4\ (aq)\ (-88\ kJ\ mol^{-1})$$

Industrially SO_3 is made by the contact process. Sulfur dioxide, generally made by the burning of sulfur or iron pyrite (a sulfide ore of iron), is first purified by electrostatic precipitation. The purified SO_2 is then oxidised by atmospheric oxygen at between 400 and 600 °C over a catalyst consisting of vanadium pentoxide (V_2O_5) activated with potassium oxide K_2O on kieselguhr or silica support. Platinum also works very well but is too expensive and is poisoned (rendered ineffective) much more easily by impurities.

The majority of sulfur trioxide made in this way is converted into sulfuric acid not by the direct addition of water, with which it forms a fine mist, but by absorption in concentrated sulfuric acid and dilution with water of the produced oleum.

3. NITROGEN OXIDES

Fig.3.3.1. Emission and deposition of nitrogen oxides and nitrate. (kodu.ut.ee)

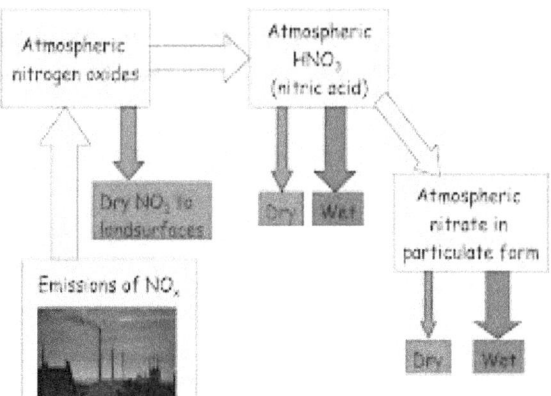

NO_x is a generic term for mono-nitrogen oxides NO and NO_2 (nitric oxide and nitrogen dioxide). They are produced from the reaction of nitrogen and oxygen gases in the air during combustion, especially at high temperatures. In areas of high motor vehicle traffic, such as in large cities, the amount of nitrogen oxides emitted into the atmosphere as air pollution can be significant. NO_x gases are formed whenever combustion occurs in the presence of nitrogen – as in an air-breathing engine; they also are produced naturally by lightning. In atmospheric chemistry, the term means the total concentration of NO and NO_2. NO_x gases react to form smog and acid rain as well as being central to the formation of tropospheric ozone. Almost all NOx emissions are in the form of NO, which has no known adverse health effects in the concentrations found in atmosphere.

However, NO can be oxidized to NO2 in the atmosphere, which in turn may give rise to secondary pollutants, which are injurious. NO2 is a reddish-brown toxic gas has a characteristic sharp, biting odor. NO_2 is one of the most prominent air pollutants. Agricultural fertilization and the use of nitrogen fixing plants also contribute to atmospheric NO_x, by promoting nitrogen fixation by microorganisms(J.N. Galloway et al. September 2004)(E.A. Davidson & W. Kingerlee 1997). Using nitrogen dioxide data acquired from 1996 to 2006 by the Global Ozone Monitoring Experiment(GOME) instrument aboard ESA's ERS-2 satellite and the aboard ESA's Envisat,Ghude was able to identify the major NO2 hotspots,quantify the trend over major industrial zones and identify the largest contributing regions. On the basis of the available data he find that nitrous oxide emission hot spots (Fig 3.3.2)correlates well with the location of mega thermal power plants,mega cities,urban and industrial regions (© Phy.org™ 2003-2013).

Figure3.3.2: Oct 22, 2007 .NO2 emission hot spots over India derived from Envisat and ERS-2 data (SCIAMACHY and GOME instruments respectively). (Indian Institute of Tropical Meteorology)

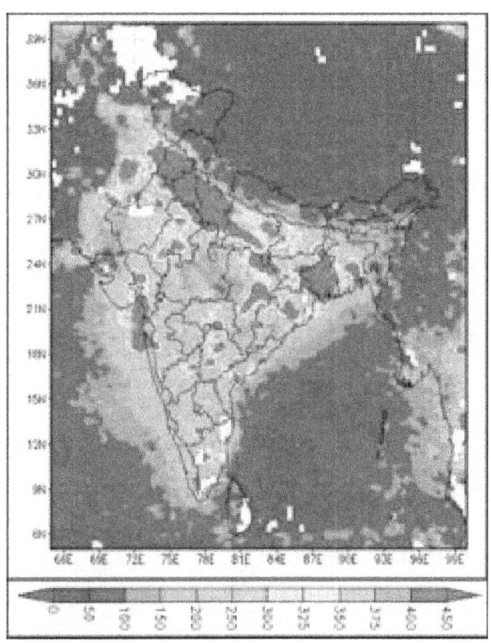

As this text is mainly concerned with anthropogenic production of pollutants so three primary sources of NO_x in combustion processes has been explained. NO_x is generated by thermal, fuel and prompt combustion.

Fig 3.3.3: Combustion Processes that generate NO_x.

Thermal NO_x formation, which is highly temperature dependent, is recognized as the most relevant source when combusting natural gas. Fuel NO_x tends to dominate during the combustion of fuels, such as coal, which have a significant nitrogen content, particularly when burned in combustors designed to minimise thermal NO_x. The contribution of prompt NO_x is normally considered negligible. A fourth source, called feed NO_x is associated with the combustion of nitrogen present in the feed material of cement rotary kilns, at between 300° and 800 °C, where it is also a minor contributor.

Thermal NO$_x$ refers to NO$_x$ formed through high temperature oxidation of the diatomic nitrogen found in combustion air (Milton R. Beychok ,March 1973). The formation rate is primarily a function of temperature and the residence time of nitrogen at that temperature. At high temperatures, usually above 1600 °C (2900 °F), molecular nitrogen (N$_2$) and oxygen (O$_2$) in the combustion air disassociate into their atomic states and participate in a series of reactions.

The three principal reactions (the extended Zeldovich mechanism) producing thermal NO$_x$ are:

$$N_2 + O \rightarrow NO + N$$
$$N + O_2 \rightarrow NO + O$$
$$N + OH \rightarrow NO + H$$

All three reactions are reversible. Zeldovich was the first to suggest the importance of the first two reactions. The last reaction of atomic nitrogen with the hydroxyl radical, OH, was added by Lavoie, Heywood and Keck to the mechanism and makes a significant contribution to the formation of thermal NO$_x$. The major source of NO$_x$ production from nitrogen-bearing fuels such as certain coals and oil, is the conversion of fuel bound nitrogen to NO$_x$ during combustion (Milton R. Beychok ,March 1973). During combustion, the nitrogen bound in the fuel is released as a free radical and ultimately forms free N$_2$, or NO. Fuel NO$_x$ can contribute as much as 50% of total emissions when combusting oil and as much as 80% when combusting coal. Although the complete mechanism is not fully understood, there are two primary paths of formation. The first involves the oxidation of volatile nitrogen species during the initial stages of combustion. During the release and prior to the oxidation of the volatiles, nitrogen reacts to form several intermediaries which are then oxidized into NO. If the volatiles evolve into a reducing atmosphere, the nitrogen evolved can readily be made to form nitrogen gas, rather than NO$_x$. The second path involves the combustion of nitrogen contained in the char matrix during the combustion of the char portion of the fuels. This reaction occurs much more slowly than the volatile phase. Only around 20% of the char nitrogen is ultimately emitted as NO_x, since much of the NO_x that forms during this process is reduced to nitrogen by the char, which is nearly pure carbon.

Prompt

This third source is attributed to the reaction of atmospheric nitrogen, N_2, with radicals such as C, CH, and CH_2 fragments derived from fuel, where this cannot be explained by either the aforementioned thermal or fuel processes. Occurring in the earliest stage of combustion, this results in the formation of fixed species of nitrogen such as NH (nitrogen monohydride), HCN (hydrogen cyanide), H_2CN (dihydrogen cyanide) and CN- (cyano radical) which can oxidize to NO. In fuels that contain nitrogen, the incidence of prompt NO_x is especially minimal and it is generally only of interest for the most exacting emission targets.

NO2 may also lead to formation of HNO3, which is washed out of the atmosphere as acid rain (Fig.3.3.4)

Fig 3.3.4: Acid rain formation.(www.elmhurst.edu)

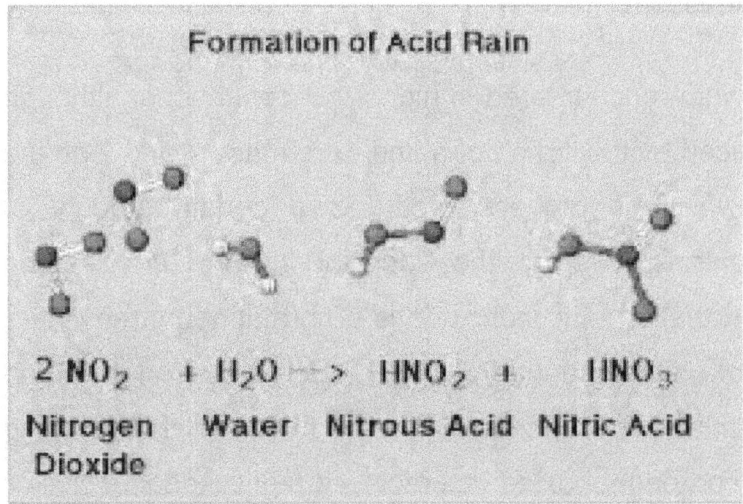

4. CARBON MONOXIDE

Carbon monoxide (CO) is a colorless, odorless, and tasteless gas that is slightly less dense than air. In the atmosphere it is spatially variable, short lived, having a role in the formation of ground-level ozone. Carbon monoxide consists of one carbon atom and one oxygen atom, connected by a triple bond that consists of two covalent bonds as well as one dative covalent bond. It is the simplest oxocarbon, and isoelectronic with the cyanide ion and molecular nitrogen. Carbon monoxide is produced from the partial oxidation of carbon-containing compounds; it forms when there is not enough oxygen to produce carbon dioxide (CO_2), such as when operating a stove or an internal combustion engine in an enclosed space. In the presence of oxygen, including atmospheric concentrations, carbon monoxide burns with a blue flame, producing carbon dioxide (Thompson, Mike, Carbon Monoxide – Molecule of the Month, Winchester College, UK). Coal gas, which was widely used before the 1960s for domestic lighting, cooking, and heating, had carbon monoxide as a significant fuel constituent. Some processes in modern technology, such as iron smelting, still produce carbon monoxide as a byproduct (Ayres, Robert U. and Ayres, Edward H. ,2009).

Worldwide, the largest source of carbon monoxide is natural in origin, due to photochemical reactions in the troposphere that generate about 5×10^{12} kilograms per year (Weinstock, B.; Niki, H. 1972). Other natural sources of CO include volcanoes, forest fires, and other forms of combustion.

It is toxic to humans and animals when encountered in higher concentrations, although carbon monoxide is naturally produced by the action of heme oxygenase 1 and 2 on the heme from hemoglobin breakdown. This process produces a certain amount of carboxyhemoglobin in normal persons, even if they do not breathe any carbon monoxide. Following the first report that carbon monoxide is a normal neurotransmitter in 1993,(Verma, et al. 1993)(Kolata, Gina .January 26, 1993) as well as one of three gases that naturally modulate inflammatory responses in the body (the other two being nitric oxide and hydrogen sulfide), carbon monoxide has received a great deal of clinical attention as a biological regulator. In many tissues, all three gases are known to act as anti-inflammatories, vasodilators, and promoters of neovascular growth (Li, L; Hsu, A;

Moore, PK .2009). Carbon monoxide is an industrial gas that has many applications in bulk chemicals manufacturing (Elschenbroich, C.; Salzer, A.2006).

Carbon monoxide is a atmospheric pollutant in urban areas, chiefly from the exhaust of internal combustion engines (including vehicles, portable and back-up generators, lawn mowers, power washers, etc.), but also from incomplete combustion of various other fuels (including wood, coal, charcoal, oil, paraffin, propane, natural gas, and trash). Most of the CO emissions are from transportation sector. Peak concentrations occur at street level in busy urban centers particularly when there is no atmospheric mixing as it happens during winter season. Thus, the largest source of CO today is motor vehicles. It has also been found a prominent indoor pollutant. In closed environments, the concentration of carbon monoxide can easily rise to lethal levels. However, according to the Florida Department of Health, "every year more than 500 Americans die from accidental exposure to carbon monoxide and thousands more across the U.S. require emergency medical care for non-fatal carbon monoxide poisoning"(Environmental Public Health Tracking – Florida Dept. of Health) .These products include malfunctioning fuel-burning appliances such as furnaces, ranges, water heaters, and gas and kerosene room heaters; engine-powered equipment such as portable generators; fireplaces; and charcoal that is burned in homes and other enclosed areas. The Centers for Disease Control and Prevention estimates that several thousand people go to hospital emergency rooms every year to be treated for carbon monoxide poisoning (Centers for Disease Control and Prevention).

Role in ground-level ozone formation

Carbon monoxide is, along with aldehydes, part of the series of cycles of chemical reactions that form photochemical smog. It reacts with hydroxyl radical ($\cdot$OH) to produce a radical intermediate $\cdot$HOCO, which transfers rapidly its radical hydrogen to O_2 to form peroxy radical ($HO_2\cdot$) and carbon dioxide (CO_2)(Reeves et al. 2002). Peroxy radical subsequently reacts with nitrogen oxide (NO) to form nitrogen dioxide (NO_2) and hydroxyl radical. NO_2 gives O (^{3}P) via photolysis, thereby forming O_3 following reaction with O_2. Since hydroxyl radical is formed during the formation of NO_2, the balance of the sequence of chemical reactions starting with carbon monoxide and leading to the formation of ozone is:

$$CO + 2O_2 + h\nu \rightarrow CO_2 + O_3$$

(where hv refers to the photon of light absorbed by the NO_2 molecule in the sequence)

Although the creation of NO_2 is the critical step leading to low level ozone formation, it also increases this ozone in another, somewhat mutually exclusive way, by reducing the quantity of NO that is available to react with ozone (Ozone and other photochemical oxidants. National Academies. 1977).

5. LEAD

Lead is a chemical element in the carbon group with symbol Pb (from Latin: plumbum) and atomic number 82. Lead is a soft and malleable metal, which is regarded as a heavy metal and poor metal. Metallic lead has a bluish-white color after being freshly cut, but it soon tarnishes to a dull grayish color when exposed to air. Lead has a shiny chrome-silver luster when it is melted into a liquid. Lead is used in building construction, lead-acid batteries, bullets and shot, weights, as part of solders, pewters, fusible alloys, and as a radiation shield. Lead, at certain contact degrees, is a poisonous substance to animals, including humans. It damages the nervous system and causes brain disorders. Lead poisoning has been documented from ancient Rome, ancient Greece, and ancient China.

The largest source of Pb in the atmosphere has been from leaded gasoline combustion, but with the gradual elimination worldwide of lead in gasoline, air Pb levels have decreased considerably (Daly, A. and P. Zannetti. 2007) .Lead released from motor vehicle exhaust may affect human populations by direct inhalation, in which case people living nearest to highways are at greatest risk. Other airborne sources include combustion of solid waste, coal, and oils, emissions from iron and steel production and lead smelters, and tobacco smoke. Lead can be ingested also after it is deposited on to foodstuffs.

Radioactive Pollutants

Radioactivity (http://epa.gov/radtown/air.htm) is an air pollutant that is both geogenic and anthropogenic(Daly, A. and P. Zannetti. 2007). Geogenic radioactivity results from the presence of radionuclides, which originate either from radioactive minerals in the earth's crust or from the interaction of cosmic radiation with atmospheric gases. Anthropogenic radioactive emissions originate from nuclear reactors, the atomic energy industry (mining and processing of reactor fuel), nuclear weapon explosions, and plants that reprocess spent reactor fuel. Since coal contains small quantities of uranium and thorium, these radioactive elements can be emitted into the atmosphere from coal-fired power plants and other sources.Radon-222 is a naturally occurring radioactive gas which comes out of ground.In the open air,it is dispersed but it can accumulate in buildings.The gas decays into minute solid particles which,if breathed in,can be deposited on the surface of the lungs and increase the risk of lung cancer (http://www.epa.gov/radtown/coal-plant.htm) .

---

6. HYDROCARBONS

Fig 3.6.1:Hydrocarbons.

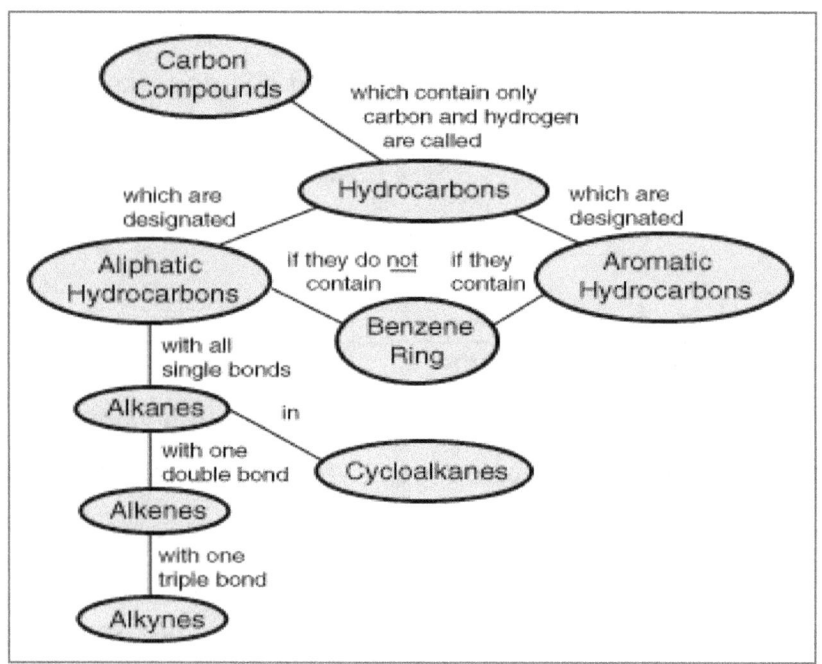

Hydrocarbon contamination in the environment is a very serious problem whether it comes from petroleum, pesticides or other toxic organic matter. One of the most significant pollutants emitted by motor vehicles in terms of quantity is hydrocarbons. The pollutant quantities of diesel particulates, tire abrasion and benzol are much smaller, but they are important because of their effects. Hydrocarbons are released through the exhaust when fuel is unburned or incompletely burned. Considerable amounts also reach the atmosphere due to fuel evaporation. Hydrocarbons evaporate from the fuel tank and other fuel feed elements, such as the fuel line, carburetor, filter, reserve canister, etc.. Hydrocarbons also vaporize when fuel station storage depots and motor vehicle tanks are filled (http://www.stadtentwicklung.berlin.de/umwelt/umweltatlas/ed 309_ 01.htm).

Environmental pollution caused by petroleum is of great concern because petroleum hydrocarbons are toxic to all forms of life. Environmental contamination by crude oil is relatively common because of its widespread use and its associated disposal operations and accidental spills. The term petroleum is referred to an extremely complex mixture of a wide variety of low and high molecular weight hydrocarbons. This complex mixture contains saturated alkanes, branched alkanes, alkenes, napthenes (homo-cyclics and hetero-cyclics), aromatics (including aromatics containing hetero atoms like sulfur, oxygen, nitrogen, and other heavy metal complexes), naptheno-aromatics, large aromatic molecules like resins, asphaltenes, and hydrocarbon containing different functional groups like carboxylic acids, ethers, etc. Crude oil also contains heavy metals and much of the heavy metal content of crude oil is associated with pyrrolic structures known as porphyrins. Petroleum is refined into various fractions such as light oil, naphtha, kerosene, diesel, lube oil waxes, and asphaltenes, etc. The light fractions, which are distilled at atmospheric pressure, are commonly known as light ends and the heavy fractions like lube oil and asphaltenes are known as the heavy ends. The light and the heavy ends of petroleum have different hydrocarbons composition, the light ends contain low molecular weight saturated hydrocarbons, unsaturated hydrocarbons, naphthenes, and low percentage of aromatic compounds; while the heavier ends consist of high molecular weight alkanes, alkenes, organometallic compounds, and high molecular weight aromatic compounds. This portion is comparatively rich in metals and N, S, O containing compounds. These hydrocarbon molecules are widespread in the environment due to the wide range of petroleum uses, which are presentedelsewhere.(http://www.dep.state.fl.us/waste/quick_topics/publications/pss/pcp/PetroleumProductDescriptions.pdf). Heavy metals are naturally present in soils; however due to human activity, the concentration of heavy metals in soil is increasing. Some areas contain such a high concentrations of heavy metals and metalloids that they are affecting the natural ecosystem. It has been observed that hydrocarbon contaminated sites co-contaminated with heavy metals are difficult to bioremediate. The reason is that heavy metals and metalloids restrict microbe's activity rendering it unable to degrade hydrocarbon or reducing its efficiency. (http://cdn.intechopen.com/pdfs/37042/InTechHydrocarbon_pollution_effects_on_living_ organisms_remediation_of_contaminated_environments_and_effects_of_heavy_metals _co_contamination_on_bioremediation.pdf).

Fig.3.6.2: Some of the representative molecules in the light and heavy fractions found in petroleum. (http://nepis.epa.gov/Adobe/PDF../P1000AE6.pdf; Wenpo et al., 2011; Dmitriev and Golovko, 2010, Qian et al., 2008)

Typical saturated linear and branched chain hydrocarbons

Benzene   Toluene   Xylene   Indene   Methyl Dibenzothiophene

Phenanthrene   Anthracene   Dibenzothiophene

Typical smaller aromatics

Polynuclear model structures

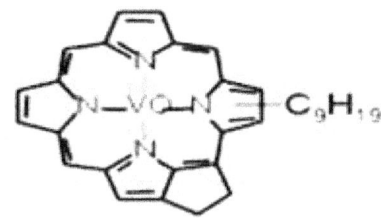

Vanadium Porphyrin

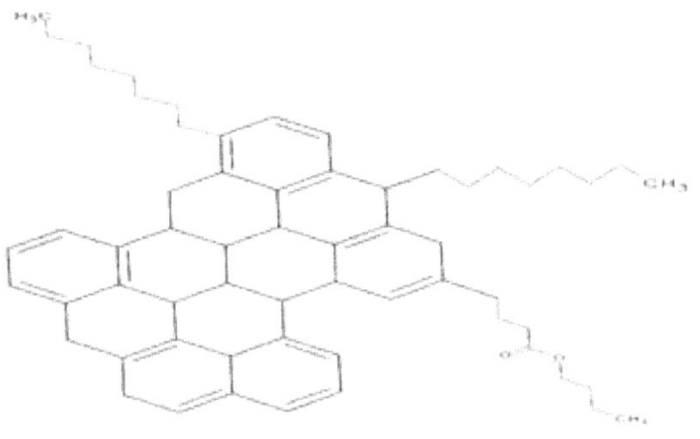

Typical heavy resin molecule

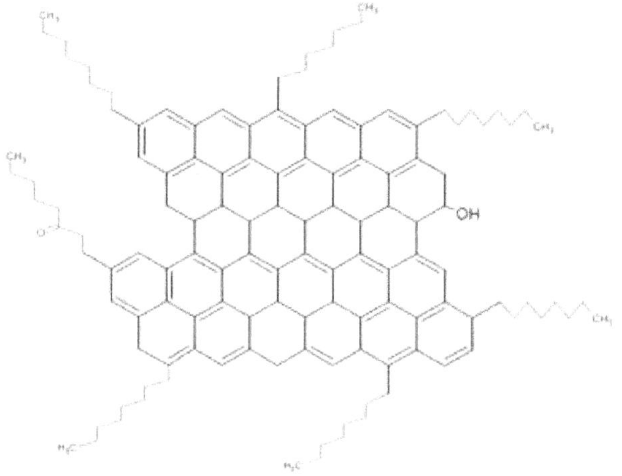

Typical asphaltene molecule

Hydrocarbons are currently the main source of the world's electric energy and heat sources (such as home heating) because of the energy produced when burnt(http://www.worldcoal.org/coal/uses-of-coal/coal-electricity/).Often this energy is used directly as heat such as in home heaters, which use either petroleum or natural gas. The hydrocarbon is burnt and the heat is used to heat water, which is then circulated. A similar principle is used to create electric energy in power plants. Burning of hydrocarbons is an example of an exothermic chemical reaction. Common properties of hydrocarbons are the facts that they produce steam, carbon dioxide and heat during combustion and that oxygen is required for combustion to take place. The simplest hydrocarbon, methane, burns as follows:

$$CH_4 + 2\ O_2 \rightarrow 2\ H_2O + CO_2 + Energy$$

In inadequate supply of air, CO gas and water vapour are formed:

$$2\ CH_4 + 3\ O_2 \rightarrow 2CO + 4H_2O$$

Another example of this property is propane:

$$C_3H_8 + 5\ O_2 \rightarrow 4\ H_2O + 3\ CO_2 + Energy$$

$$C_nH_{2n+2} + (3n+1)/2\ O_2 \rightarrow (n+1)\ H_2O + n\ CO_2 + Energy$$

In order to study the oxidation of hydrocarbons, the complex engine system is divided into two parts(CPCB.November 2010): (a) the combustion chamber and (b) the exhaust system. Inside the cylinder a fraction of hydrocarbons is oxidized during the expansion and exhaust process and the products are carbon dioxide, carbon monoxide and ―hydrocarbons containing oxygen‖ like Aldehydes, Ketones, Ethers and Alcohols etc. The general mechanism for the formation of Hydrocarbon (HC) is that fuel or fuel-air mixture escapes the primary combustion process to the exhaust system. This fuel (containing HC) then survives the expansion processes and passes through the exhaust system without oxidation and end up in the atmosphere as HC emissions. With the use of catalysts or a thermal reactor in the exhaust system, substantial reductions in HC emission levels can be achieved. The exhaust system temperature might be sufficient for partial oxidation, but not high enough for complete combustion. Cylinder-left-out gases are subjected to oxidation in the exhaust manifold. The engine parameters that influence on oxidation process are speed, spark timing, mixing composition, compression ratio, air / fuel ratio, load and heat losses (J.B.Heywood, 1988).

Factors which influence the engine conditions on hydrocarbon emissions are as follows(CPCB,2010):

(a) Fuel/Air Ratio: HC emission increases rapidly as the fuel and air mixture becomes fuel rich (Slone et al (1989), Quader et al (1989), Grimm et al (1988). For example, the hydrocarbon mass emitted under rich condition (Φ =1.15) was twice that of lean condition (Φ =0.90). The relative abundance of methane and acetylene increased during fuel-rich operation. The presence of excess oxygen or excess hydrocarbons during combustion has a significant effect on the relative concentration of the various oxygenated hydrocarbon species emitted. As the availability of oxygen increases, the emission of hydrocarbon partial oxidizing products increases.

(b) Temperature: The combustion rate of hydrocarbons increases rapidly with temperature. Fundamental studies using combustion bombs and spectroscopic measurement . of excited species in flames have shown that the combustion rate of hydrocarbons increases rapidly with temperature (Slone et al., 1989). In general, olefins produce higher flame temperatures than aromatics and both are higher than paraffin's (Quader et al., 1989). Spark retardation leads to reduced peak temperature but increased temperature late in the cycle and in the exhaust. NOx decreases because of the lower peak temperature while HCs decrease because of the increased burn-up resulting from higher late cycle and exhaust temperature. An increase in engine speed usually results in increased combustion chamber temperature and increased catalyst temperature depending upon the thermal heat transfer properties of the particular engine and exhaust system.

Fuel parameters that influence hydrocarbon emissions are as follows:

(a) Fuel Molecular weight: The studies of single cylinder engines revealed that the total engine-out emission is increased as the average molecular weight of the fuel increased.

(b) Fuel Volatility: The change in the distillation temperature can change the engine exhaust gas. In some of the fuels, the decrease in HC emissions (approximately 23%) due to change in there distillation temperature from 360ºC to 138 oC. The higher HC emissions observed using a higher T90 fuel were probably because of the combined effects of the increased absorption of the heavier hydrocarbons ion oils films, on metal surfaces and in cylinder deposits (Siegl et al .1992, Kaiser et al., 1991).

(c) Paraffin content in fuel: When simple paraffin's are substituted for olefins in the fuel, the atmospheric reactivity of exhaust pipe emissions decreases and the non-methane hydrocarbon increases.

(d) Aromatic content in fuel: A decrease in aromatics and olefins and replacement by paraffins will result in an increased production of molecular hydrogen during combustion. The H/C ratio of fuel aromatics and olefins are lower than those of the normal and iso-paraffins. The decrease in aromatics and/or olefins helps in reduced HC emissions, assuming all other important variables remain constant.

(e) Oxygenated compound in fuel: The addition of oxygenated fuel components has the effect of leaning –out the air/fuel ratio, which results in reduced HC and CO emissions.

Presently, only NOx (Oxides of Nitrogen), CO (Carbon Monoxide), THC (Total Unburned Hydrocarbons) particulates and visible smoke are those emissions which are legislated for the respective type of fueled vehicles. In addition to ongoing regulated parameters, the impact of increasing traffic density on the environment and public health demands need the limits for Aldehydes, Ketones and Methane emissions. A study on these parameters in vehicle/ engine exhaust provides inputs for emission regulations thereby providing scope for reducing pollution levels and consequently improving public health in the country.

Formaldehyde and Acetaldehyde are the most common non methane Carbonyl compounds found in vehicular exhaust.

(i) Formaldehyde is formed due to incomplete combustion of both gasoline and diesel fuel and accounts for 1 to 4% of Total Organic Gas (TOG) emissions, depending on control technology and fuel consumption.

(ii) Vehicle emissions do not only contribute primary formaldehyde but also responsible formaldehyde formed from photo oxidation of the VOC.

(iii) Acetaldehyde is a saturated aldehyde found in vehicle exhaust due to incomplete combustion of fuel.

(iv) Acetaldehyde constitute 0.4 to 1 % of exhaust total organic Gas (TOG), depending on control technology and fuel composition.

(v) Mobile sources contribute to both primary and secondary emissions of acetaldehyde.

(vi) Aldehydes get involved in chemical reaction in the atmosphere, generating other compounds, some of which leads to photochemical smog formation that mostly produces oxidizing gases, especially ozone gas.

(vii) Oxidizing gases formed due to chemical reaction in atmosphere also damage materials like rubber.

Process chemistry Release/Mechanism.

Aldehyde and Ketone formation

Aldehydes and ketones found in exhaust gases are formed in the engine and exhaust system. In the basic aldehyde formation reactions, the important carriers of the chain reaction in the combustion of hydrocarbons are the alkyl radicals (R), which are formed by cleavages of C-C or C-H bonds of hydrocarbons. At high temperatures, dehydrogenation by oxygen and presence of other combustion related radicals influence the cleavages processes.

Regarding the higher aldehydes, the general mechanism is based on the formation of the alkyl radicals (RO_2.)

$$R + O_2 = RO_2. \quad \quad (1)$$

This alkyl radical is the precursor for the further reactions, including intra-molecular hydrogen abstractions and decomposition. Finally leading to the following aldehyde formation reactions:

$$RCH_2OO. \; RC. \; HOOH \; RCHO + .OH \quad \quad (2)$$

$$RCH_2O + O_2 \rightarrow RCHO + HO_2. \quad \text{(3)}$$

$$RCH=CH_2 + \cdot O \rightarrow RCHO \quad \text{(4)}$$

An ambient temperature range in an engine exhaust system of 300-2000 K, the temperature has the large effect on the aldehyde formation.

Aldehydes are formed by oxidation during the combustion and by oxidation of unburnt hydrocarbons. Sources of unburnt hydrocarbons are given below (J.B.Haywood, 1988).

Flame quenching

Flame quenching at the cool combustion chamber walls, which results in a thin layer of unburned fuel/air mixture close to the wall after flame passage.

Crevices mechanism

Crevices in the combustion chamber wall which are too narrow for flame to enter leads to the fuel/air mixture escaping from the primary combustion processes.

Absorption and desorption of fuel vapor in the oil layer and deposits

The oil layer present on the combustion chamber wall and deposits formed on the combustion chamber absorb fuel vapor during intake and compression processes, and this fuel is desorbed during the expansion and exhaust processes.

Gas-phase quenching

It has been created when the engine is operating under extreme conditions of equivalence ratio and spark timing.

Leakage of unburnt mixture through the valve

Valve leakage can occur which leads to a small fraction of fuel/air mixture escaping the primary combustion processes.

Unevaporated liquid fuel in the cylinder

For liquid fuels, an important process which could contribute to hydrocarbon emissions is the liquid fuel within the cylinder which fails to evaporate and mix with sufficient air to burn before the end of combustion, particularly during the engine starting and warm-up process.

The ketone formation also starts from the alkyl-peroxy radicals RO_2 (Miller et al, 1990). The ketone reaction equation as follows:

RR'CHOO. RR'C.OOH RR'CO+.OH --------------------------------- (5)

2 Methane Formation

Methane is emitted from light duty vehicles due to the incomplete combustion of fuel from the engine and the incomplete oxidation of engine-out methane in current catalytic after treatment systems. It is important to recognize, that current vehicles produce and emit substantially less methane than their older counterparts.

3 Carbonyl Emission

The selective carbonyl and methane emission literature study has been done on the engines using various fuels. The literature given below is the resources which have been used many times in the study of carbonyl emission on engines using various fuels. Very less studies are being available in the public domain.

Gasoline fueled engines

Numerous studies have been conducted on gasoline fueled engine for analyzing carbonyl emissions.

Yao et al. (2008) suggested that reduction in aromatic content in Gasoline increases aldehyde emission. Karl-Erik Egebäck et al (2005) said that, the emissions of aldehydes (especially acetaldehyde and formaldehyde) from vehicles running on ethanol/gasoline blends are expected to increase. Wigg et al, (1973) reported that increasing the aromatic content in fuel reduces formaldehyde emissions from the exhaust. Oberdorfer et al, (1967) denoted that, in gasoline the presence of different groups like (Aromatics, Naphthene, Paraffin, Olefins) have different effect of carbonyl emissions (aldehyde & ketone). The increasing order of carbonyl emission based on the groups is given below:

Aromatic < Naphthene < Paraffin < Olefin

Methane emission:

Increases in ethanol percentage in gasoline enhances the methane emission in the exhaust may be due to increase in octane number of the gasoline which has the probability of knocking (Anonymous, 2008). Due to the incomplete combustion inside the engine and the incomplete oxidation of methane at exhaust develop higher methane emission (RE Hayes, 2004).

Properties of Aldehydes, Ketones and Methane:

Aldehydes and ketones are partially oxygenated organic compounds containing carbonyl group. An Aldehyde functional group consists of a carbon atom bonded to a hydrogen atom and double-bonded to an oxygen atom (O=CH-). Whereas a ketone functional group contains a carbonyl group (C=O) bonded to two other carbon atoms.

Table 3.6.1. The physiochemical properties of the most common carbonyl compounds found in engine exhaust emissions are given in table. (CPCB.November 2010)

Number	IUPAC name	Synonym	C No.	Formula	Molecular Weight kg/mol	Density kg/m3	Melting Point °C	Boiling point °C
1	Methanal	Aldehyde Formalydehyde	1	HCHO	30.03	815	-92	-21
2	Ethanal	Acetaldehyde	2	CH_3CHO	44.05	778	-123	20
3	Propanal	Propionaldehyde	3	CH_3CH_2CHO	58.08	797	-81	48
4	Butanal	n-Butylraldehyde	4	$CH_3(CH_2)_2CHCHO$	72.11	803	-97	75
5	Pentanal	n-Valeraldehyde	5	$CH_3(CH_2)_3CHO$	86.13	808	-91	103
6	Hexanal	n-capronaldehyde	6	$CH_3(CH_2)_4CHO$	100.16	814	-56	128
7	Propenal	Acrolein	3	CH_2CHCHO	56.07	841	-86	53
8	Trans-2-Butenal	Crotonaldehyde	4	$CH_3CHCHCHO$	70.09	852	-74	102
9	2Methyl 2propenal	Methacrolein	4	CH_2CCH_3CHO	70.09	843	-81	68

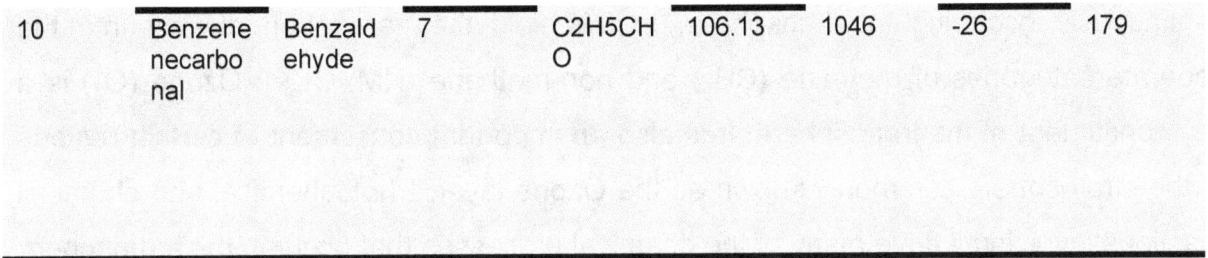

| 10 | Benzenecarbonal | Benzaldehyde | 7 | C2H5CHO | 106.13 | 1046 | -26 | 179 |

7. GROUND LEVEL OZONE (O₃)

Fig3.7.1 : (a) Ground level ozone and stratospheric ozone.(b) Ozone formation

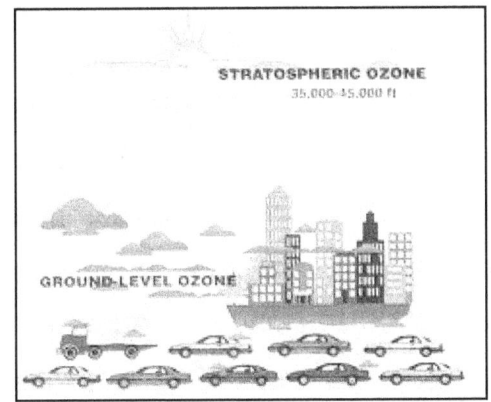

(a) (www.wmcac.org)   *(b) (airnow.gov)*

Ozone is found in two regions of the Earth's atmosphere – at ground level and in the upper regions of the atmosphere(fig. 3.7.1,a). Both types of ozone have the same chemical composition (O3). While upper atmospheric ozone protects the earth from the sun's harmful rays, ground level ozone is the main component of smog(EPA).

Tropospheric, or ground level ozone, is not emitted directly into the air, but is a secondary pollutant formed when sunlight causes photochemical reactions involving NOX and VOCs(fig.3.7.1,b). VOCs are an important outdoor air pollutant. These are thousands of such compounds. The most anthropogenic source is the automobile. They may be converted into harmful compounds through complex chemical reactions continuously occurring in atmosphere. In this field they are often divided into the separate categories of methane (CH_4) and non-methane (NMVOCs). Ozone (O_3) is a key constituent of the troposphere. It is also an important constituent of certain regions of the stratosphere commonly known as the Ozone layer. Photochemical and chemical reactions involving it drive many of the chemical processes that occur in the atmosphere by day and by night. At abnormally high concentrations brought about by human activities (largely the combustion of fossil fuel), it is a pollutant, and a constituent of smog. Ozone concentrations tend to peak in the afternoon. Ozone is likely to reach unhealthy levels on hot sunny days in urban environments. Ozone can also be transported long distances by wind. For this reason, even rural areas can experience high ozone levels.

Emissions from industrial facilities and electric utilities, motor vehicle exhaust, gasoline vapors, and chemical solvents are some of the major sources of NOx and VOC and hence cause of ground level ozone. Under the Clean Air Act, EPA has established health and environmentally protective standards for ozone in the air . EPA and others have instituted a variety of multi-faceted programs to meet these standards. Additional programs are being put into place in many countries to cut NOx and VOC emissions from vehicles, industrial facilities, and electric utilities. Programs are also aimed at reducing pollution by reformulating fuels and consumer/commercial products, such as paints and chemical solvents that contain VOC. Voluntary and innovative programs encourage communities to adopt practices, such as carpooling, to reduce harmful emissions.

8. HYDROGEN SULFIDE.

It has specific, easily identifiable rotten-egg odor with formula H_2S. it is heavier than air, very poisonous, corrosive, flammable and explosive. Hydrogen sulfide often results from the bacterial breakdown of organic matter in the absence of oxygen, such as in swamps and sewers; this process is commonly known as anaerobic digestion. H_2S also occurs in volcanic gases (via the hydrolysis of sulfide minerals, i.e. $MS + H2O \rightarrow MO + H$

2S), natural gas, biogas, LPG and some well waters. The human body produces small amounts of H$_2$S and uses it as a signaling molecule. Dissolved in water, hydrogen sulfide is known as hydrosulfuric acid or sulfhydric acid, a weak acid. Hydrogen sulfide and oxygen burn with a blue flame to form sulfur dioxide (SO2) and water. In general, hydrogen sulfide acts as a reducing agent. At high temperature or in the presence of catalysts, sulfur dioxide can be made to react with hydrogen sulfide to form elemental sulfur and water. This is exploited in the Claus process, the main way to convert hydrogen sulfide into elemental sulfur. If gaseous hydrogen sulfide is put into contact with concentrated nitric acid, it explodes. Hydrogen sulfide reacts with alcohols to form thiols, an important class of organosulfur compounds.

Small amounts of hydrogen sulfide occur in crude petroleum, but natural gas can contain up to 90%. By far the largest industrial route to H$_2$S occurs in petroleum refineries: The hydrodesulfurization process liberates sulfur from petroleum by the action of hydrogen. The resulting H$_2$S is converted to elemental sulfur by partial combustion via the Claus process. Other anthropogenic sources of hydrogen sulfide include coke ovens, paper mills (using the sulfate method), and tanneries. H$_2$S arises from virtually anywhere where elemental sulfur comes in contact with organic material, especially at high temperatures.

Hydrogen sulfide is a highly toxic and flammable gas (flammable range: 4.3–46%). Being heavier than air, it tends to accumulate at the bottom of poorly ventilated spaces. Although very pungent at first, it quickly deadens the sense of smell, so potential victims may be unaware of its presence until it is too late. For safe handling procedures, a hydrogen sulfide material safety data sheet (MSDS) should be consulted (Iowa State University, Department of Chemistry MSDS. "Hydrogen Sulfide Material Safety Data Sheet").

9. MERCURY

Mercury is a naturally occurring element (Hg on the periodic table) that is found in air, water and soil. It exists in several forms: elemental or metallic mercury, inorganic mercury compounds, and organic mercury compounds. Elemental or metallic mercury is a shiny, silver-white metal and is liquid at room temperature. Pure mercury is a liquid metal, sometimes referred to as quicksilver that volatizes readily. If heated, it is a colorless, odorless gas. Exposures to mercury can affect the human nervous system and harm the brain, heart, kidneys, lungs, and immune system (http://www.epa.gov/mercury/about.htm). It has traditionally been used to make products like thermometers, switches, and some light bulbs. Mercury is found in many rocks including coal. When coal is burned, mercury is released into the environment. Coal-burning power plants are the largest human-caused source of mercury emissions to the air in the United States, accounting for over 50 percent of all domestic human-caused mercury emissions (Source: 2005 National Emissions Inventory). EPA has estimated that about one quarter of U.S. emissions from coal-burning power plants are deposited within the contiguous U.S. and the remainder enters the global cycle. Burning hazardous wastes, producing chlorine, breaking mercury products, and spilling mercury, as well as the improper treatment and disposal of products or wastes containing mercury, can also release it into the environment. Mercury in the air eventually settles into water or onto land where it can be washed into water. Once deposited, certain microorganisms can change it into methylmercury, a highly toxic form that builds up in fish, shellfish and animals that eat fish. Fish and shellfish are the main sources of methylmercury exposure to humans. Methylmercury builds up more in some types of fish and shellfish than others. The levels of methylmercury in fish and shellfish depend on what they eat, how long they live and how high they are in the food chain. Another exposure to mercury that can be a concern is breathing mercury vapor. These exposures can occur when elemental mercury or products that contain elemental mercury break and release mercury to the air, particularly in warm or poorly-ventilated indoor spaces (http://www.epa.gov/mercury/about.htm).

10. CADMIUM

Cadmium is a soft silver-white metal that is usually found in combination with other elements. Cadmium compounds range in solubility in water from quite soluble to practically insoluble (U.S. Department of Health and Human Services, Atlanta, GA. 1997). The chemical symbol for cadmium is Cd and the atomic weight is 112.41

g/mol(U.S. Department of Health and Human Services, Atlanta, GA. 1997). Cadmium is used to manufacture pigments and batteries and in the metal-plating and plastics industries. The largest sources of airborne cadmium in the environment are the burning of fossil fuels such as coal or oil, and incineration of municipal waste materials. Cadmium may also be emitted into the air from zinc, lead, or copper smelters(U.S. Department of Health and Human Services, Atlanta, GA. 1997) . Smoking is another important source of cadmium exposure. Smokers have about twice as much cadmium in their bodies as do nonsmokers. EPA has classified cadmium as a Group B1, probable human carcinogen.

11. CHROMIUM

The metal, chromium (Cr), is a steel-gray solid with a high melting point and an atomic weight of 51.996 g/mol. Chromium has oxidation states ranging from chromium (-II) to chromium (+VI). Chromium forms a large number of compounds, in both the chromium (III) and the chromium (VI) forms. Chromium compounds are stable in the trivalent state, with the hexavalent form being the second most stable state. The chromium (III) compounds are sparingly soluble in water and may be found in water bodies as soluble chromium (III) complexes, while the chromium (VI) compounds are readily soluble in water(U.S. Department of Health and Human Services, Atlanta, GA. 1998). Chromium is a naturally occurring element in rocks, animals, plants, soil, and volcanic dust and gases. The metal chromium is used mainly for making steel and other alloys. Chromium compounds, in either the chromium (III) or chromium (VI) forms, are used for chrome plating, the manufacture of dyes and pigments, leather and wood preservation, and treatment of cooling tower water. Smaller amounts are used in drilling muds, textiles, and toner for copying machines. Chromium occurs in the environment predominantly in one of two valence states: trivalent chromium (Cr III), which occurs naturally and is an essential nutrient, and hexavalent chromium (Cr VI), which, along with the less common metallic chromium (Cr 0), is most commonly produced by industrial processes. (U.S. Department of Health and Human Services, Atlanta, GA. 1998) .Air emissions of chromium are predominantly of trivalent chromium, and in the form of small particles or aerosols. (U.S. Department of Health and Human Services, Atlanta, GA. 1998)(

SAIC. 1998.).The most important industrial sources of chromium in the atmosphere are those related to ferrochrome production. Ore refining, chemical and refractory processing, cement-producing plants, automobile brake lining and catalytic converters for automobiles, leather tanneries, and chrome pigments also contribute to the atmospheric burden of chromium(EPA.1988). The general population is exposed to chromium (generally chromium [III]) by eating food, drinking water, and inhaling air that contains the chemical. Dermal exposure to chromium may occur during the use of consumer products that contain chromium, such as wood treated with copper dichromate or leather tanned with chromic sulfate. Occupational exposure to chromium occurs from chromate production, stainless-steel production, chrome plating, and working in tanning industries; occupational exposure can be two orders of magnitude higher than exposure to the general population. People who live in the vicinity of chromium waste disposal sites or chromium manufacturing and processing plants have a greater probability of elevated chromium exposure than the general population. These exposures are generally to mixed chromium (VI) and chromium (III). (U.S. Department of Health and Human Services, Atlanta, GA. 1998).

12. ASBESTOS

Asbestos is the name applied to a group of six different minerals that occur naturally in the environment. The most common mineral type is white, but others may be blue, gray, or brown. These minerals are made up of long, thin fibers that are somewhat similar to fiberglass. Asbestos is neither volatile nor soluble; however, small fibers may occur in suspension in both air and water. The main uses of asbestos are in building materials, paper products, asbestos-cement products, friction products, textiles, packings and gaskets, and asbestos-reinforced plastics (U.S. Public Health Service, U.S. Department of Health and Human Services, Atlanta, GA. 1995).EPA has classified asbestos as a Group A, known human carcinogen.Airborne exposure to asbestos may occur through the erosion of natural deposits in asbestos-bearing rocks, from a variety of asbestos-related industries, or from clutches and brakes on cars and trucks. The concentrations in outdoor air are highly variable. Asbestos has been detected in indoor air, where it is released from a variety of building materials such as insulation and ceiling and floor tiles.

13. DIOXIN AND FURANS

Fig 3.13.1: Dioxins and Furans. *(pubs.usgs.gov)*

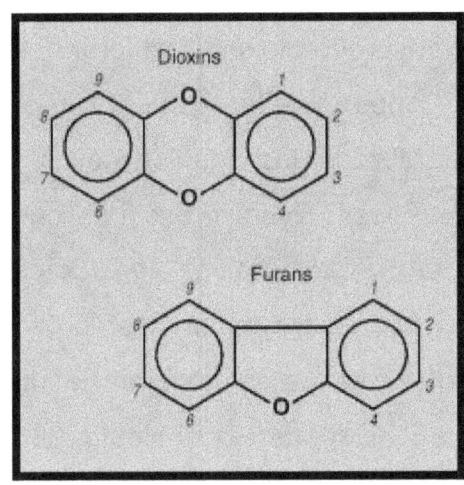

Dioxins and furans is the abbreviated or short name for a family of toxic substances that all share a similar chemical structure. Dioxins, in their purest form, look like crystals or a colorless solid. Most dioxins and furans are not man-made or produced intentionally, but are created when other chemicals or products are made. Of all of the dioxins and furans, one, 2,3,7,8-tetrachloro-p-dibenzo-dioxin (2,3,7,8 TCDD) is considered the most toxic.(Agency for Toxic Substances and Disease Registry (ATSDR).1999). Dioxins and furans are not made for any specific purpose; however, they are created when products like herbicides are made. In many developing countries Persistent Organic Pollutants (POPs) are still used for agricultural and disease vector control, as well as for industrial purposes. So, the stockpiles of obsolete POPs create significant problems that are compounded by municipal-waste burning on open sites. When this waste is burned at low temperatures, it produces significant quantities of polychlorinated dioxins and related chemicals. Surveys, conducted in a number of Asian countries, indicate that such activities lead to local, regional and widespread global contamination. Studies show that alarming levels of POPs are present in the environment as well as in human beings. (Scientific and Technical Advisory Panel.2004) They are also created in the pulp and paper industry, from a process that bleaches the wood pulp. In addition, they can be produced when products are burned. Dioxins and furans can enter our body through breathing contaminated air, drinking contaminated water or eating contaminated food. About 90% of exposure to dioxins and furans is from eating contaminated food(http://www.chiefs-ofontario.org/eagle/factsheet11.htm).Dioxins and furans can build up in the fatty tissues of animals.We can be exposed to dioxins and furans by eating contaminated food. Dioxins and furans typically stay and build up in the fatty tissues of animals. This means that eating beef, pork, poultry, fish as well as dairy products can be a source of exposure. Furan is found in heat-treated commercial foods and it is produced through thermal degradation of natural food constituents. Notably, it can be found in roasted coffee, instant coffee, and processed baby foods.(Moro, S.2012) (European Food Safety Authority .2011). There are several sources of exposure to dioxins and furans. If you work in or near a municipal solid waste incinerator, copper smelter, cement kiln or coal fired power plant you can be exposed to dioxins and furans. Individuals who burn their household waste or burn wood can be exposed as well. Even forest fires can contribute to the creation of small amounts of dioxins and furans.Dioxins and furans have been found in the air,soil, and food. Dioxins and furans are mainly distributed through the air. However, only a small percentage of exposure is from air.Eating contaminated food is the primary source of exposure. U.S. (Department of Health and Human Services. 2001)

Dioxins and furans are some of the most toxic chemicals known to science. A draft report released for public comment in September 1994 by the US Environmental Protection Agency clearly describes dioxin as a serious public health threat. The public health impact of dioxin may rival the impact that DDT had on public health . Dioxin was the primary toxic component of Agent Orange, was found at Love Canal in Niagara Falls, NY and was the basis for evacuations at Times Beach, MO and Seveso, Italy.

The **International Council of Local Environmental Initiatives (ICLEI)-South Asia** undertook a project to learn the **'Energy and Carbon Emissions Profiles of 54 South Asian Cities'** during 2007-08, and collected relevant data with the cooperation of the individual cities' Urban Local Bodies (ULBs) and their utilities.

The report hence produced in 2009 provides an inventory of energy consumption and carbon emissions data of 54 South Asian cities, which include 41 cities from India, 4 cities each from Sri Lanka and Bangladesh, 3 from Nepal and 2 from Bhutan. For each of the 54 selected cities, sector-wise emissions data was summarized.

It is well known that carbon emissions have played a key role in what has come to be known as *the global warming phenomena*. Global warming is caused by human activities, which alter the chemical composition of the atmosphere through a build-up of greenhouse gases – primarily carbon dioxide, methane and nitrous oxide, and particulate matter.

If these emissions remain unchecked, the steadily rising global temperatures will cause sea-levels to rise, alter local weather conditions considerably, affect forests, crop yields and water supplies. They would also affect human health, flora and fauna, and alter ecosystems permanently.

Scientific evidence shows that global warming is responsible for environmental changes that can result in irreversible climate change – *extreme* and *erratic weather events* that can severely impact life.

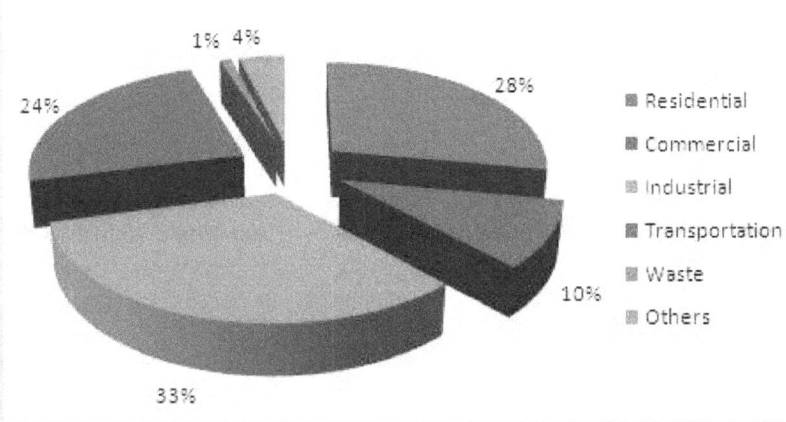

Fig 1:Sector –wise Emissions in the Selected Indian Cities,2007-2008. (ICLEI)

Top 10 Carbon Emitting Cities of India as found by ICLEI :

1. Kolkata (West Bengal)

With a total carbon emission of 9.33 Million Tonnes for 2007-08, Kolkata becomes the highest carbon emitting city of India.

2. Visakhapatnam (Andhra Pradesh)

For the duration 2007-08, Visakhapatnam recorded 7.36 Million Tonnes eCO2 and was the second highest carbon emitting cities of India.

3. Ahmedabad (Gujarat)

With its carbon emissions reaching 6.78 Million Tonnes eCO2 during 2007-08, the city of Ahmedabad finds itself named as the third highest emitter in India.

4. Bangalore (Karnataka)

Once a lush green, coconut canopied city with pleasant weather, Bangalore became the "Silicon Valley of India and emitted 6.36 Million Tonnes eCO2 during 2007-08 to became the 4th highest carbon emitter in India.

5. Pune (Maharashtra)

With an estimated 6.00 Million Tonnes eCO2 emissions in 2007-08, Pune – or Poona as it was called during British rule – stands as the fifth largest carbon emitter in India.

6. Jamshedpur (Jharkhand)

Jamshedpur's carbon emissions amounted to 5.51 Million Tonnes eCO2 in 2007-08, and make it the sixth highest emitter in India. Jamshedpur also has the highest per capita carbon emissions of all the 41 cities.

7. Chennai (Tamil Nadu)

Chennai, or Madras as it was known formerly, showed 3.82 Million Tonnes eCO2 for the duration 2007-08, and became the seventh highest carbon emitting cities in India.

8. Surat (Gujarat)

Once the site of plague epidemic in 1994, causing a number of countries to impose temporary travel and trade sanctions, Surat is listed at 8th position amongst India's top carbon emitting cities, and recorded 3.38 Million Tonnes eCO2 emissions during 2007-08.

9. Ranchi (Jharkhand)

Ranchi, known for pleasant climate, water-falls and having a hill-station-like status only three decades ago, reported carbon emissions of 2.88 Million Tonnes eCO2 during 2007-08, and acquired ninth place amongst highest carbon emitting cities of India.

10. Gurgaon (Haryana)

Vicinity to national capital New Delhi and resulting confluence of businesses and migrant population has converted Gurgaon from a farming village to a bustling city listed as the tenth highest carbon emitting cities and recording

2.55 Million Tonnes eCO2 of emissions in 2007-08.

Table 2: Carbon Emissions in 41 Indian Cities, 2007-08 (in Million Tonnes).

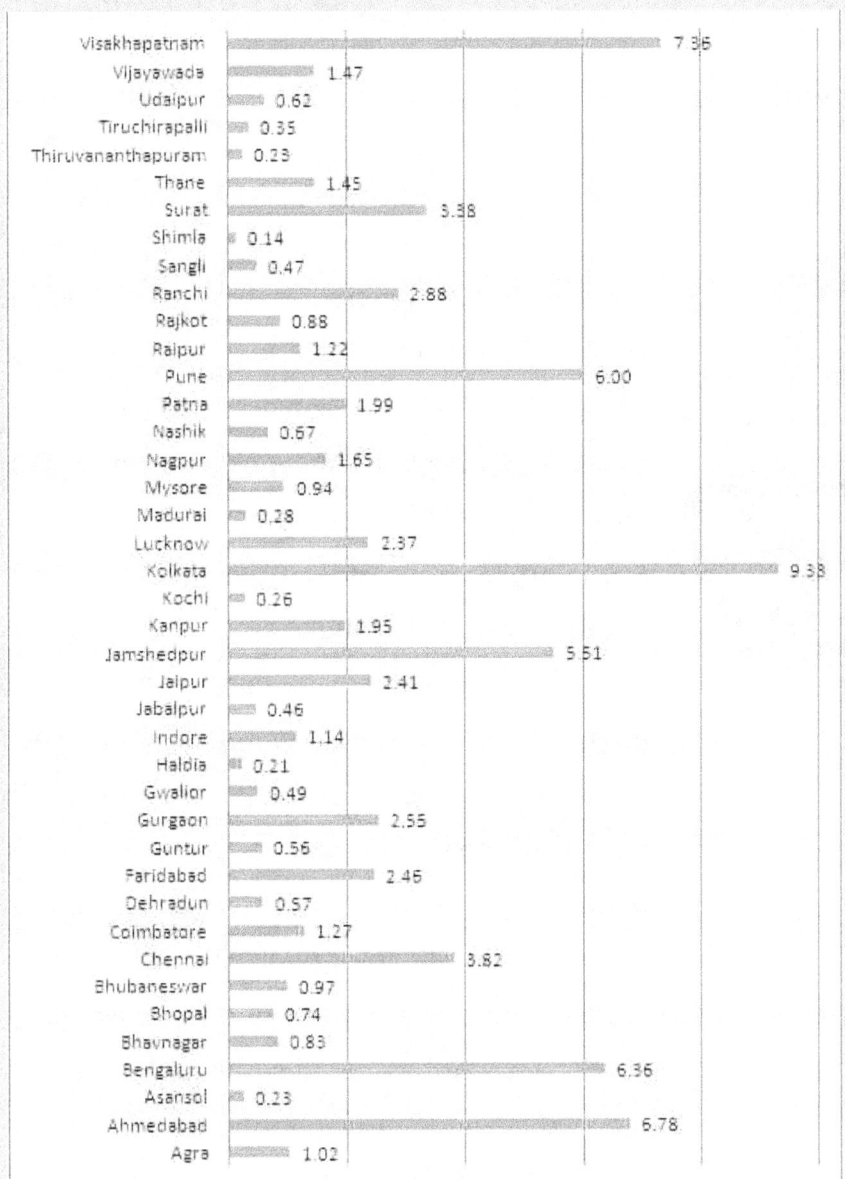

(http://marketspace.thinktosustain.com/2012/07/top-10-carbon-emitting-cities-in-india/#.UfDM_GTn_Dc)

4. Effects of Pollutants

Fig4.1:Effects of Pollution. (www.epa.gov)

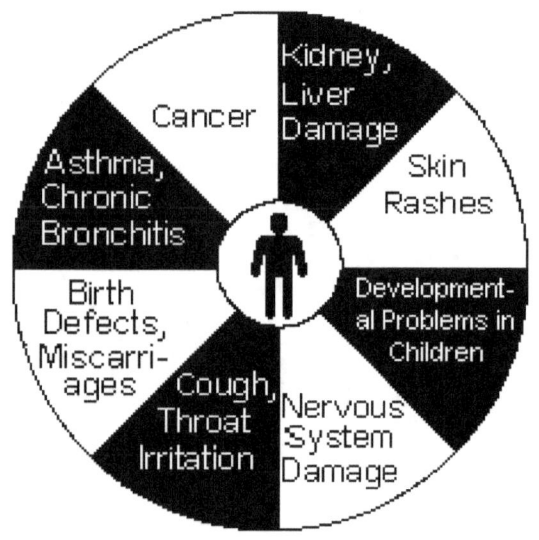

Air pollution is one of the several environmental factors that are having a serious impact on human health and quality of life. Air pollution is a significant risk factor for multiple health conditions including respiratory infections, heart disease, and lung cancer, according to the WHO. The health effects caused by air pollution may include difficulty in breathing, wheezing, coughing, asthma and aggravation of existing respiratory and cardiac conditions. These effects can result in increased medication use, increased doctor or emergency room visits, more hospital admissions and premature death. The human health effects of poor air quality are far reaching, but principally affect the body's respiratory system and the cardiovascular system. Individual reactions to air pollutants depend on the type of pollutant a person is exposed to, the degree of exposure, the individual's health status and genetics. Both indoor and outdoor air pollution have caused approximately 3.3 million deaths worldwide. Children aged less than five years that live in developing countries are the most vulnerable population in terms of total deaths attributable to indoor and outdoor air pollution("Air quality and health". www.who.int.). The World Health Organization states that 2.4 million people die each year from causes directly attributable to air pollution, with 1.5 million of these deaths attributable to indoor air pollution("Estimated deaths & DALYs attributable to selected environmental risk factors, by WHO Member State, 2002"). Around the world, children living in cities with high exposure to air pollutants are at increased risk of developing asthma, pneumonia and other lower respiratory infections. Because children are outdoors more and have higher minute ventilation they are more susceptible to the dangers of air pollution. Risks of low initial birth weight are also heightened in such cities. The World Health Organization reports that the greatest concentrations of particulates are found in countries with low economic world power and high poverty and population growth rates. What is even more alarming is that the World Health Organization (WHO) estimates that by the year 2020 there will be eight million deaths related to air pollution alone (Cifuentes et al.2001). The situation is particularly grim in developing countries such as India where deaths due to pollution exceed those due to cancer and heart disease combined(Bradford G.Hill et al.2006).Protective measures to ensure children's health are being taken in cities such as New Delhi, India where buses now use compressed natural gas to help eliminate the "pea-soup" smog(World Health Organization).

Projected Premature Annual Deaths due to Urban Air Pollution, Total and by Economic Group or Region, 2001–2020

Region	Premature Deaths (thousand per year)
Established market economies	20
Former socialist economies	200
China	590
India	460
East Asia and the Pacific	150
Latin America and the Caribbean	130
South Asia	120
Middle East Crescent	90
Sub-Saharan Africa	60
World	1,810

Source: World Bank.

A study by the University of Birmingham has shown a strong correlation between pneumonia related deaths and air pollution from motor vehicles (*The Guardian*.2008-04-15). Worldwide more deaths per year are linked to air pollution than to automobile accidents (Collins, Nick .April 18, 2012). A 2005 study by the European Commission calculated that air pollution reduces life expectancy by an average of almost nine months across the European Union (BBC. February 21, 2005). Causes of deaths include aggravated asthma, emphysema, lung and heart diseases, and respiratory allergies (US Department of Energy).

Fig.4.3:Airway infection and inflammation.(Manas Ranjan Ray , Twisha Lahiri 2010)

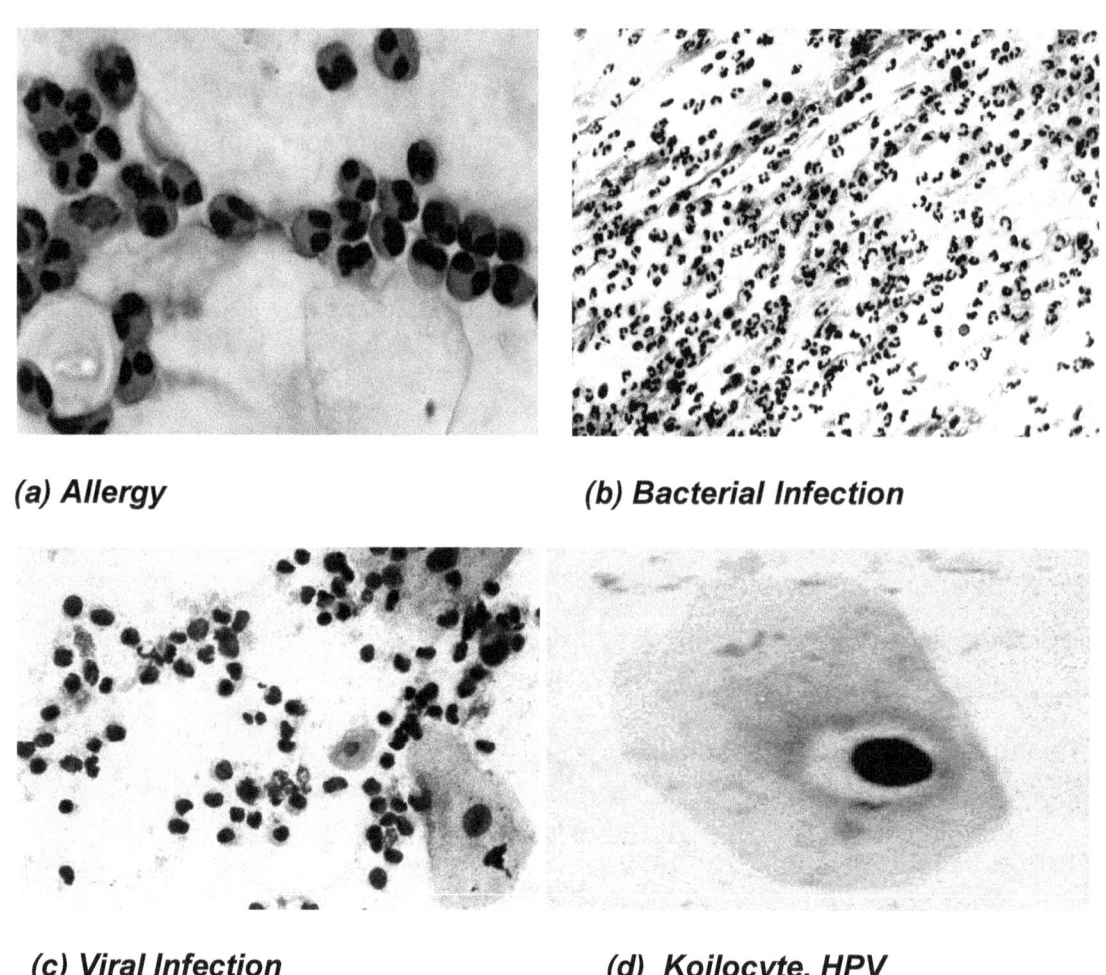

(a) Allergy

(b) Bacterial Infection

(c) Viral Infection

(d) Koilocyte, HPV

In December 1993, Harvard researchers published the results of a sixteen-year-long community health study that tracked the health of 8,000 adults in six U.S. cities with differing levels of air pollution. After adjusting for age and smoking, researchers found that residents of the most polluted city had a 26 percent higher mortality rate than those living in the least polluted city. This translated into a one- to two-year shorter lifespan for residents of the most polluted cities. Another major study corroborated these findings. The study correlated American Cancer Society data on the health of 1.2 million adults with air pollution data in 151 U.S. metropolitan areas. The study found that people living in the most polluted area had a 17 percent greater risk of mortality than people living in the least polluted city. A number of prestigious international panels, including a British Committee on the Medical Effects of Air Pollutants and a Committee of the Health Council of the Netherlands, have concluded that there is a cause-and-effect relationship between particulate pollution and premature death. Such a conclusion is warranted based on the consistency of the association in different studies and situations, the dose-response relationship, and the biological plausibility.In 1996, U.S. EPA published a risk assessment focusing on Southeast Los Angeles County. The U.S. EPA estimates over 3,000 excess deaths occur annually due to levels of particle pollution above the current federal standards in this particular area of Los Angeles alone. The federal agency estimated more than 52,000 episodes of respiratory symptoms each year-including about 1,000 hospital admissions-from the particle levels observed in 1995 in Southeast Los Angeles. U.S. EPA estimates more than 40,000 particle-related health effects (including 300 to 700 deaths) would occur in Los Angeles even if the area brought pollution down to the current federal particle standards. NRDC performed a study entitled *Breath Taking: Premature Mortality Due to Particulate Air Pollution in 239 American Cities,* which was based on the risk relationships identified in the American Cancer Society and Harvard studies. In this study, released in May, 1996, NRDC applied the known risk relationships to a variety of urban areas where particle levels had been adequately monitored. They found that nationally over 50,000 premature deaths per year may be attributable to the existing levels of particles in the air.

Our past has seen catastrophic effects of pollutants . The worst short term civilian pollution crisis in India was the 1984 Bhopal Disaster (Simi Chakrabarti. "20th anniversary of world's worst industrial disaster". Australian Broadcasting Corporation) .Leaked industrial vapours from the Union Carbide factory, belonging to Union Carbide, Inc., U.S.A., killed more than 25,000 people outright and injured anywhere from 150,000 to 600,000. The United Kingdom suffered its worst air pollution event when the December 4 Great Smog of 1952 formed over London. In six days more than 4,000 died, and 8,000 more died within the following months. An accidental leak of anthrax spores from a biological warfare laboratory in the former USSR in 1979 near Sverdlovsk is believed to have been the cause of hundreds of civilian deaths. The worst single incident of air pollution to occur in the US occurred in Donora, Pennsylvania in late October, 1948, when 20 people died and over 7,000 were injured (Davis; Devra .2002).

Fig 4.4 : Auto Rikshaw over loaded with school going children and emmiting so much pollution-Daily story in India.

Even in the areas with relatively low levels of air pollution, public health effects can be significant and costly, since a large number of people breathe in such pollutants. In India condition is rather grave. According to SAARC air pollution is already a serious and widespread problem in India. India has a sizable air pollution control programme but it too faces several gaps and weaknesses in its programme. Here millions of people breathe air with high concentrations of dreaded pollutants. The air is highly polluted in terms of suspended particulate matter in most cities. This has led to a greater incidence of associated health effects on the population manifested in the form of sub-clinical effects, impaired pulmonary functions, use of medication, reduced physical performance, frequent medical consultations and hospital admissions with complicated morbidity and even death in the exposed population..A survey by the Central Pollution Control Board and the All India Institute Of Medical Sciences of New Delhi showed that a majority of people living in Delhi suffered from eye irritation, cough, sore throat, shortness of breath and poor lung functioning. One in 10 people have asthma in Delhi (CPCB.2006). Scientists have concluded that growing up in a city with polluted air is about as harmful to a person's health as growing up with a parent who smokes. Although air pollution is concentrated in cities, it can move well beyond them: for example, acidic lakes in Scandinavia have been linked to pollution from factories in the United States.

Pollutants in air is now classified into different groups depending upon their severity of effects.

Hazardous air pollutants (HAPs) "may reasonably be anticipated to be carcinogenic, mutagenic (Clean Air Act Amendments 1990), and exhibit other adverse health effects" (U.S. Clean Air Act). People exposed to toxic air pollutants at sufficient concentrations and durations may have an increased chance of getting cancer or experiencing other serious health effects. These health effects can include damage to the immune system, as well as neurological, reproductive (e.g., reduced fertility), developmental, respiratory and other health problems. In addition to exposure from breathing air toxics, some toxic air pollutants such as mercury, dioxins/furans and PAHs, can deposit onto soils or surface waters, where they are taken up by plants and ingested by animals and are eventually magnified up through the food chain(Butler et al. 1993; Ramesh et al. 2004; U.S. EPA 2003) . Like humans, animals may experience health problems if exposed to sufficient quantities of air toxics over time (http://www.epa.gov/air/toxicair/newtoxics.html).

Effects on cardiovascular health

A 2007 review of evidence found ambient air pollution exposure is a risk factor correlating with increased total mortality from cardiovascular events (range: 12% to 14% per a 10 microg/m3 increase)(PMID 19235364).Air pollution is also emerging as a risk factor for stroke, particularly in developing countries where pollutant levels are highest (Farrah J. Mateen & Robert D. Brook2011). A 2007 study found that in women air pollution is associated not with hemorrhagic but with ischemic stroke   (Miller K. A et al .2007). Air pollution was also found to be associated with increased incidence and mortality from coronary stroke in a cohort study in 2011(Andersen et all.2011). Associations are believed to be causal and effects may be mediated by vasoconstriction, low-grade inflammation or autonomic nervous system imbalance or other mechanisms (Brook et al .2010)(Louwies et al.2013). A study conducted by CNCI Kolkata found that people live in polluted urban area have higher risk of cardiovascular disease than rural people(Manas Ranjan Ray and Twisha Lahiri "Air pollution and its Effects on Health-Case Studies,India")(fig 4.4).

Fig4.4: Inflammation and activation of coagulation cascade. Increased formation of leukocyte-platelet aggregates causes thrombotic disorders and leads to CVD. (Manas Ranjan Ray, Twisha Lahiri .2010)

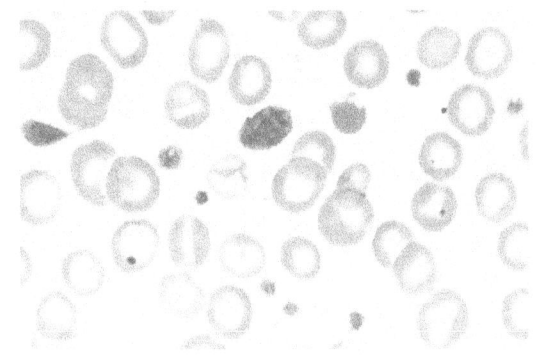

a. **Giant platelets**

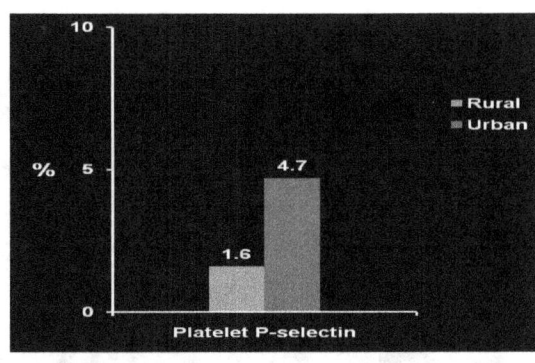

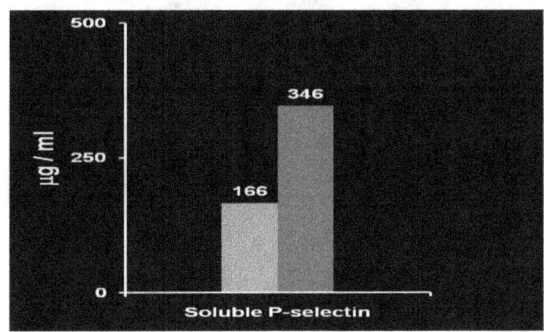

b. Air pollution activates blood platelets

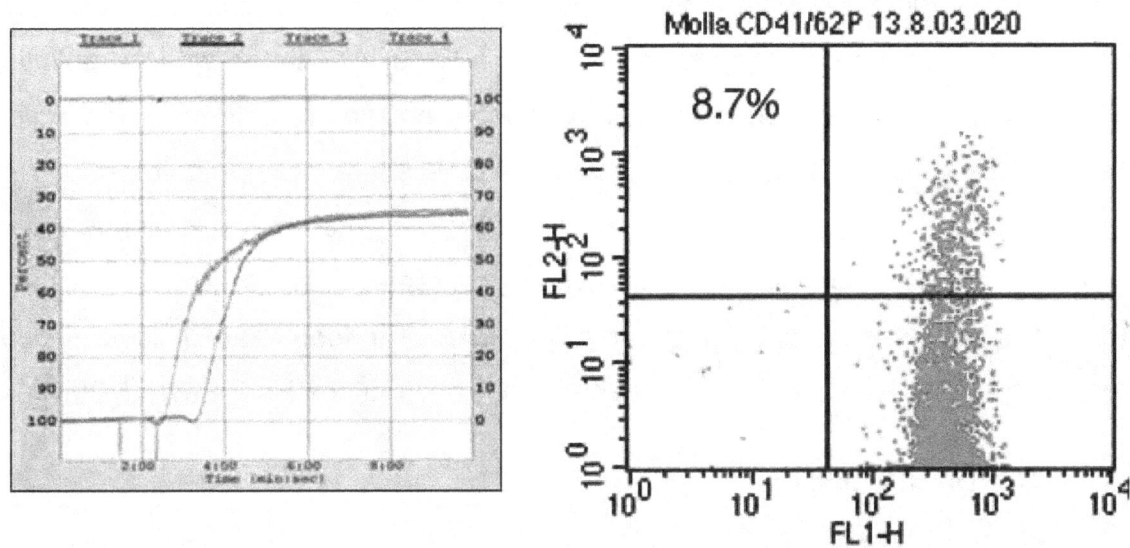

c. *Platelet activation – cardiovascular risk 2-fold rise in aggregation & ATP-release*

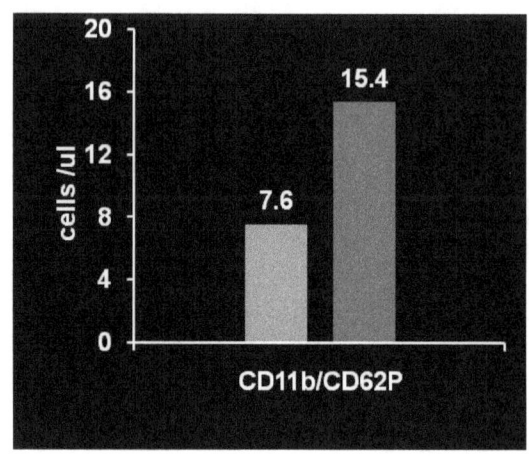 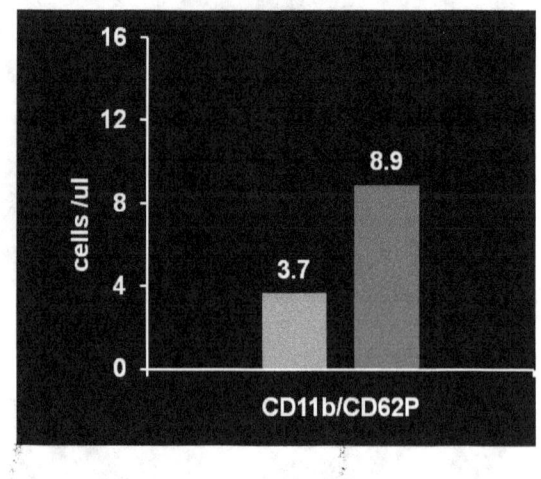

d. Neutrophile-platelets aggregates. aggregates

e. Monocyte –platelet

Effects on cystic fibrosis

A study from around the years of 1999 to 2000, by the University of Washington, showed that patients near and around particulates air pollution had an increased risk of pulmonary exacerbations and decrease in lung function (Christopher H. Goss et al.2004). Patients were examined before the study for amounts of specific pollutants like *Pseudomonas aeruginosa* or *Burkholderia cenocepacia* as well as their socioeconomic standing. During the time of the study 117 deaths were associated with air pollution. Many patients in the study lived in or near large metropolitan areas in order to be close to medical help. These same patients had higher level of pollutants found in their system because of more emissions in larger cities. As cystic fibrosis patients already suffer from decreased lung function, everyday pollutants such as smoke, emissions from automobiles, tobacco smoke and improper use of indoor heating devices could further compromise lung function(Michael Kymisis, Konstantinos Hadjistavrou .2008).

Effects on COPD and asthma

Fig 4.5 : COPD.(www.nhlbi.nih.gov)

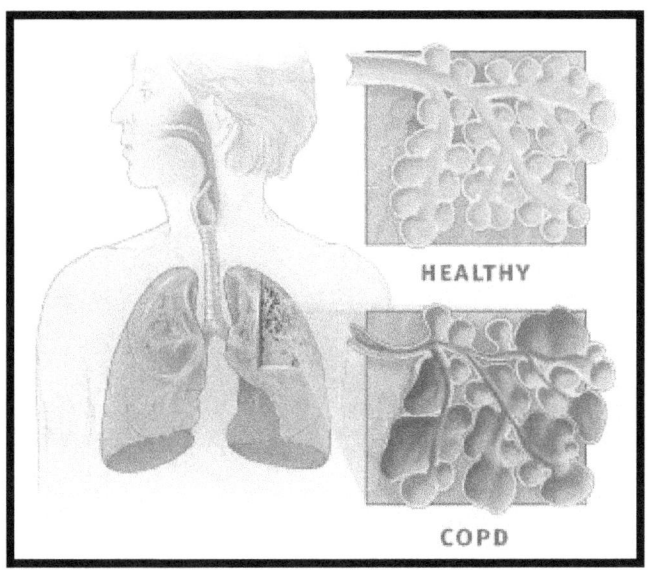

Chronic obstructive pulmonary disease (COPD) includes diseases such as chronic bronchitis and emphysema (Zoidis, John D. (1999). Researches have demonstrated increased risk of developing asthma (Gehring,U. 2010) and COPD(Andersen,Z.J.2011) from increased exposure to traffic-related air pollution. Additionally, air pollution has been associated with increased hosptializations and mortality from asthma and COPD. (Committee of the Environmental and Occupational Health Assembly of the American Thoracic Society. (1996)(Andersen et al.2011). Recent reviews conclude that near-roadway traffic emissions may not only trigger asthma symptoms, but also contribute to the development of asthma in children (Anderson et al. 2010a;

Anderson et al. 2010b). It has been found that a substantial proportion of asthma-related morbidity in chidren is a consequence of near roadway pollution, even if symptoms are triggered by other factors (http://dx.doi.org/10.1289/ehp.1104785).It is believed that much like cystic fibrosis, by living in a more urban environment serious health hazards become more apparent.

Fig 4.6: Percentage of COPD individuals in rural and urban west Bengal. (Manas Ranjan Ray and Twisha Lahiri.2010) .

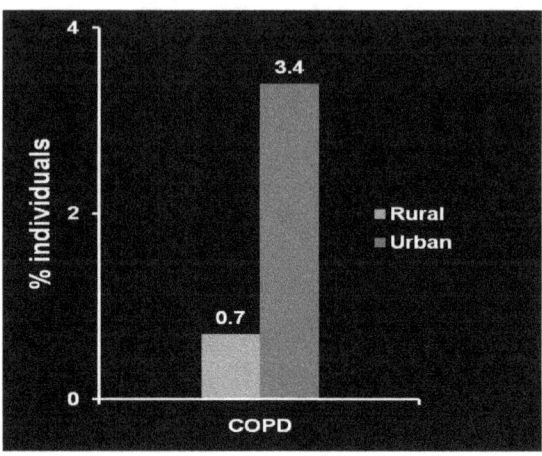

Effect on child birth weight

In the last decade, numerous studies reported associations between levels of ambient air pollutants and adverse birth outcomes (Kannan et al. 2006), although results are not consistent regarding the relevance of specific pollutants or the trimester of exposure. Associations between particulate matter (PM) and pregnancy outcomes differ by study, although many findings suggest an association. Exposure to PM10 and PM2.5 in aerodynamic diameter during gestation has been associated with LBW in some studies (Huynh et al. 2006; Morello-Frosch et al. 2010; Seo et al. 2010), but not others (Madsen et al. 2010; Seo et al. 2010). More studies have been conducted for gaseous pollutants, although results have also been inconsistent, such as for nitrogen dioxide (NO2) (Maroziene et al. 2002; Morello-Frosch et al. 2010), sulfur dioxide (SO2) (Bobak et al. 1999; Lin et al. 2004), carbon monoxide (CO) (Liu et al. 2003; Ritz et al. 1999), and ozone (O3) (Morello-Frosch et al. 2010; Salam et al. 2005). Several literature reviews of pollutant effects on adverse birth outcomes have noted that results are heterogeneous across studies, but have nonetheless concluded that associations between air pollution and adverse pregnancy outcomes are likely causal (Maisonet et al. 2004; Sapkota et al. 2010; Shah et al. 2011). Shah et al. concluded that exposure to PM2.5 is likely associated with LBW, preterm birth, and small for gestational age births (Shah et al. 2011). These reviews noted that further studies are necessary to clarify which pollutants are the most harmful and to identify during which periods of pregnancy infants are most vulnerable. Inconsistencies among previous studies might result from differences in study populations or in study design such as control for confounders, exposure assessment, statistical methods, and sample size. Other possible explanations are variation in the exposure period and collinearity among pollutants (Maisonet et al. 2004). However, a key reason studies on PM and pregnancy outcomes differ is that the chemical composition of particles varies by location (Bell et al.2007a). National Institute of Environmental Health and Sciences (NIEHS,U.S)performed astudy to explore the association between PM2.5 chemical composition and pregnancy outcomes. Chemical components of aluminum,calcium, elemental carbon, nickel, silicon, titanium and zinc were identified as a potentially harmful.It identified associations between birth outcomes and multiple PM2.5 chemical components. For gaseous pollutants, LBW was associated with exposure to CO, NO2, and SO2. Results also indicated a negative association between O3 and LBW. Some of these results (i.e. CO,NO2, SO2) are similar to those from previous studies (Darrow et al. 2011,Wu et al. 2011). Associations between LBW and individual pollutants differed by trimester. Higher exposure of specific pollutants in the first trimester may relate to placenta development, whereas exposure in later stages may affect maternal vascular alternation, which causes the fetal growth retardation (Lin et al. 1999, Mannes et al. 2005). They found statistically significant associations

with LBW for exposure during the first trimester to PM2.5 aluminum, elemental carbon, and titanium, second trimester for PM2.5 aluminum, and for exposure during the third trimester to PM10, PM2.5 aluminum, calcium, nickel, silicon and zinc. The biological mechanisms that may contribute to effects of air pollution on birth outcomes are uncertain, and various hypotheses exist. For instance, NO2 exposure during pregnancy may limit placental vascular function and disturb fetal growth (Clifton et al. 2001). CO may react with oxygen on hemoglobin binding sites, reducing oxygen delivery (Maisonet et al. 2004). Fetal growth may be retarded by direct toxic effects of air pollution, similar to effects of smoking (Ritz et al. 1999). The mechanism of PM effects on birth outcomes could be related to the transfer of toxic components to the fetus from PM that has accumulated in the mother's lungs (Ritz et al. 2007). PM has a complex chemical composition, and its chemical components may affect outcomes through different biological pathways. One possible explanation is that exposure to PM2.5 metal related components, including aluminum and titanium, increases oxidative stress burdens leading to adverse health outcomes (Wei et al. 2009) (http://dx.doi.org/10.1289/ehp.1104763).

Links to Cancer

Cancers are primarily an environmental disease with 90-95% of cases attributed to environmental factors and 5-10% due to genetics(WHO.2009). *Environmental*, as used by cancer researchers, means any cause that is not genetic. Common environmental factors that contribute to cancer death include environmental pollutants and tobacco (25-30%), diet and obesity (30-35%), infections (15-20%), radiation (both ionizing and non-ionizing, up to 10%), stress, lack of physical activity (WHO, 2009). In 2008 approximately 12.7 million cancers were diagnosed (excluding non-melanoma skin cancers and other non-invasive cancers) and 7.6 million people died of cancer worldwide(Bedard PL et al.2007). Cancers as a group account for approximately 13% of all deaths each year with the most common being: lung cancer (1.3 million deaths), stomach cancer (803,000 deaths), colorectal cancer (639,000 deaths), liver cancer (610,000 deaths), and breast cancer (519,000 deaths)(WHO.2006,Retrieved on 2011). This makes invasive cancer the leading cause of death in the developed world and the second leading cause of death in the developing world(Bedard PL et al.2007). Over half of cases occur in the developing world (Bedard PL et al.2007). According to a data collected from PGIMS, Rohtak (Haryana) India, deaths due to cancer has found to be increasing from 1991 to 2000 (Fig.4.8). In 2011, a large Danish epidemiological study found an increased risk of lung cancer for patients who lived in areas with high nitrogen oxide concentrations. In this study, the association was higher for non-smokers than smokers(Raaschou-Nielsen, O.2011). An additional Danish study, also in 2011, likewise noted evidence of possible associations between air pollution and other forms of cancer, including cervical cancer and brain cancer (Raaschou-Nielsen, O.2011) .

Fig 4.7 : Cancer mortality in men and women .(Jemal A, Siegel R, Ward E et al. (2008). "Cancer statistics, 2008". CA Cancer J Clin 58 (2): 71–96. doi:10.3322/CA.2007.0010. PMID 18287387)

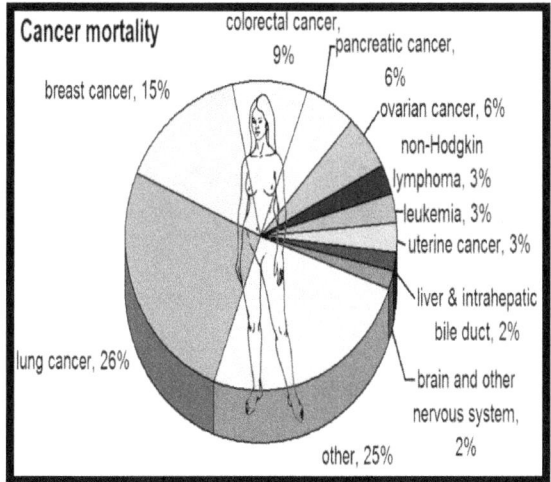
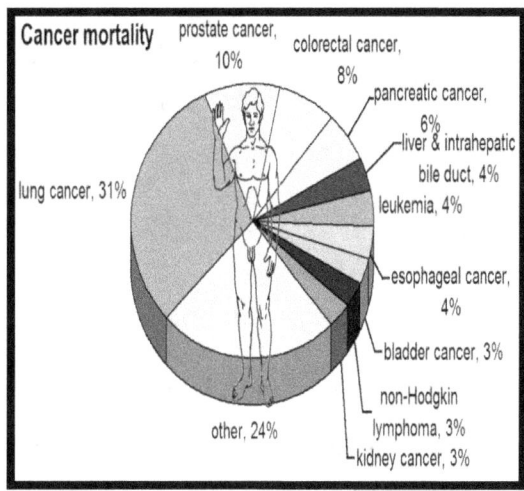

Fig 4.8 :Cancer deaths in PGIMS Rohtak hospital between 1991 – 2000 .(Data from central registration office PGIMS Rohtak (Haryana) India)

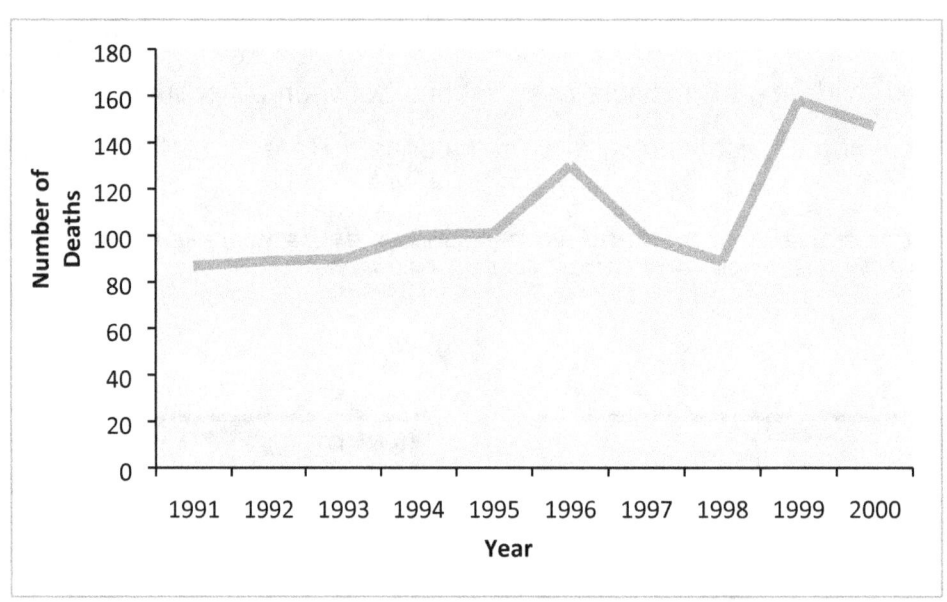

Reproductive and Developmental effects

Women exposed to high level of air pollution while pregnant were up to twice as likely to have a child with autism as women who lived in areas with low pollution,according to a new study from Harvard School of Public Health(HSPH)(The Tribune.August,2013).The researchers examined data from Nurses "Health Study II,a long term study based at Brigham and Women's Hospital involving 116,430 nurses that began in 1989. Among that group ,the authors studied 325 women who had a child with autism and 22000 women who had a child without the disorder. The results showed that women who lived in the 20% of locations with the highest levels of Diesel particulates or mercury in the air were twice as likely to have a child with autism as those who lived in the 20% of areas with the lowest levels. Others types of air pollution – lead, manganese ,methylene, chloride and combined metals exposure / were associated with higer autism risk as well. Most pollutants were associated with autism more strongly in boys than girls .

Formidable Air Pollution and Children

Air pollution is a major environment-related health threat to children and a risk factor for both acute and chronic respiratory disease(http://www.who.int/ceh/risks/cehair/en/index.html). While second-hand tobacco smoke and certain outdoor pollutants are known risk factors for respiratory infections, indoor air pollution from solid fuels is one of the major contributors to the global burden of disease. In poorly ventilated dwellings, indoor smoke can be 100 times higher than acceptable levels for small particles. Exposure is particularly high among women and young children, who spend the most time near the domestic hearth.

Outdoor air pollution is large and increasing a consequence of the inefficient combustion of fuels for transport, power generation and other human activities like home heating and cooking. Combustion processes produce a complex mixture of pollutants that comprises of both primary emissions, such as diesel soot particles and lead, and the products of atmospheric transformation, such as ozone and sulfate particles.Urban outdoor air pollution is estimated to cause 1.3 million deaths worldwide per year (http://www.who.int/ceh/risks/cehair/en/index.html). Children are particularly at risk due to the immaturity of their respiratory organ systems. Those living in middle-income countries disproportionately experience this burden.

Indoor cooking and heating with biomass fuels (agricultural residues, dung, straw, wood) or coal produces high levels of indoor smoke that contains a variety of health-damaging pollutants. There is consistent evidence that exposure to indoor air pollution can lead to acute lower respiratory infections in children under age five, and chronic obstructive pulmonary disease and lung cancer in adults.

Indoor air pollution is responsible for 2 million deaths annually (http://www.who.int/ceh/risks/cehair/en/index.html). Acute lower respiratory infections, in particular pneumonia, continue to be the biggest killer of young children and this toll almost exclusively falls on children in developing countries.

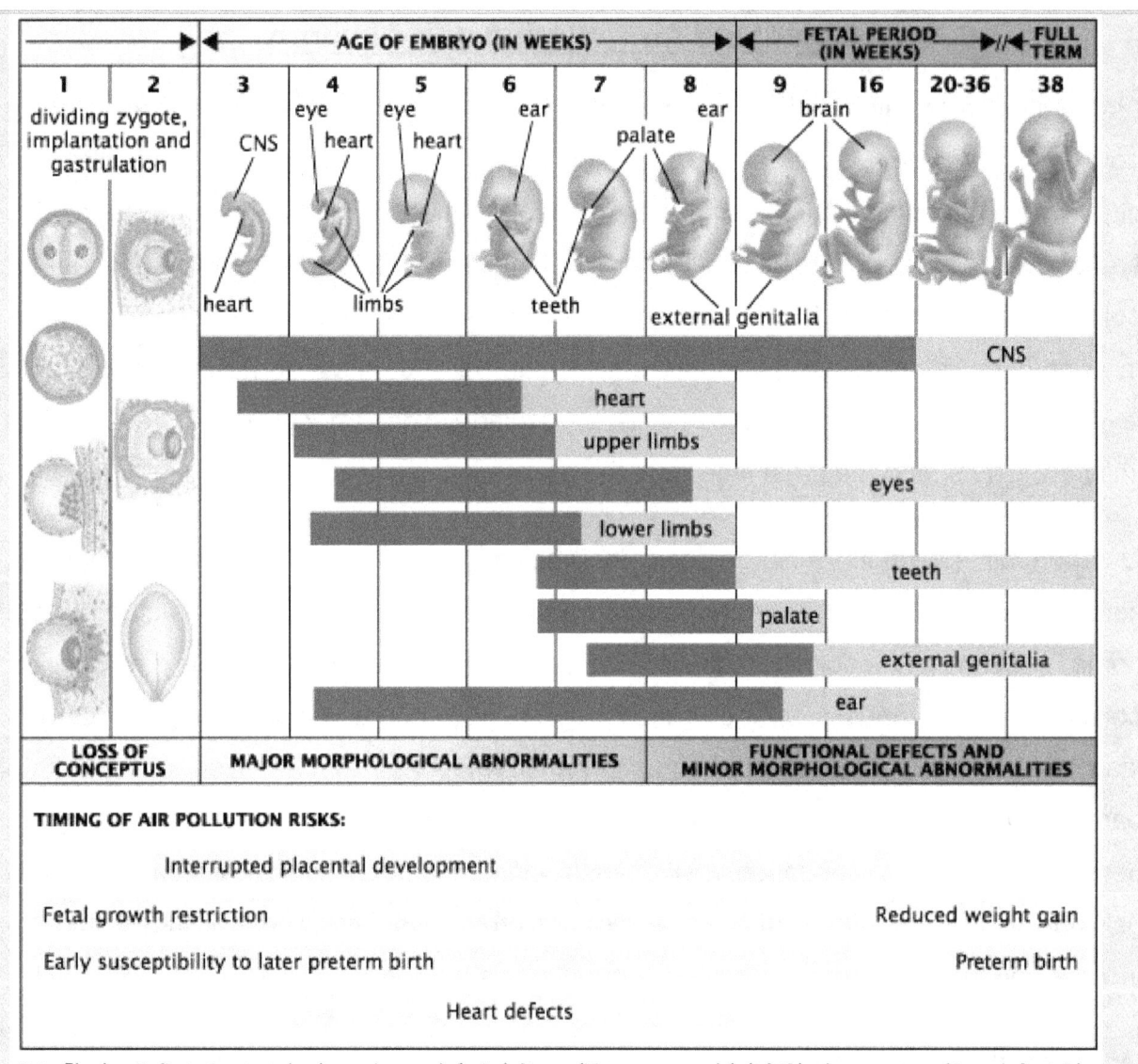

Figure 1. Fetal development and timing of air pollution risks.(Beate Ritz and Michelle Wilhelm . 2008)

Before taking Birth

The time between conception and birth is perhaps one of the most vulnerable life stages, during which the environment may have tremendous immediate and lasting effects on health (Beate Ritz and Michelle Wilhelm . 2008). The fetus undergoes rapid growth and organ development and the maternal environment helps direct these processes, for better or for worse (Figure 1). Evidence is accumulating that environmental exposures can cause infants to be born premature (before 37 weeks of gestation) or low weight (less than 2500 grams, or 5.5 pounds), or to be born with certain birth defects. These babies are far more likely to die in infancy, and those who survive have high risks of brain, respiratory, and digestive problems in early life. The impact of environmental exposures on fetal development may be far-reaching, as data suggest growth and developmental delays *in utero* influence the risk for heart disease and diabetes in adulthood.

Early childhood is also a critical period for the continued development and maturation of several biological systems such as the brain, lung, and immune system and air toxics can impair lung function and neurodevelopment, or exacerbate existing conditions, such as asthma (Figure 2). Infants who were born premature or growth-retarded may be particularly vulnerable to additional environmental insults, for example, due to immaturity of the lungs at birth(Beate Ritz and Michelle Wilhelm . 2008).

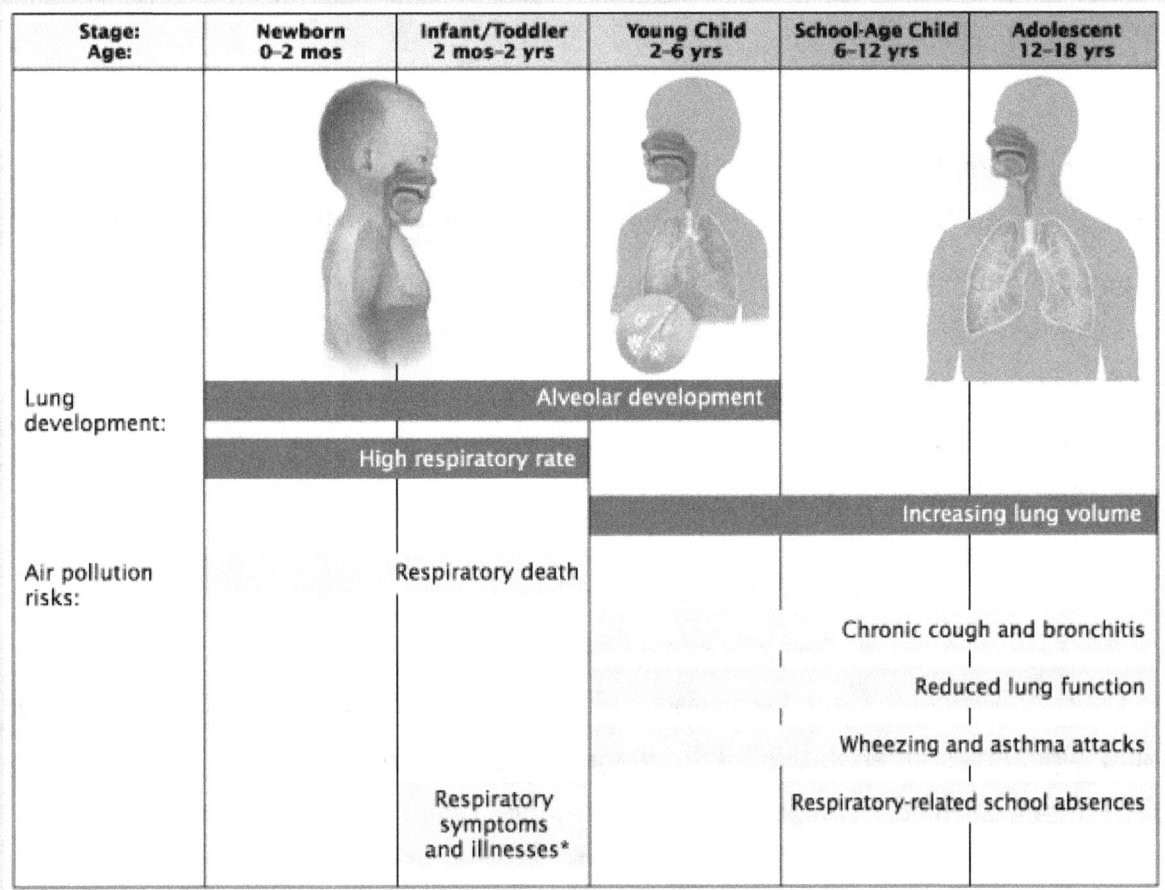

*Air pollution exposure has also been more recently linked to respiratory symptoms and illnesses in early life including cough, bronchitis, wheeze and ear infections

Figure 2. Air pollution effects on the developing respiratory system. *(Beate Ritz and Michelle Wilhelm . 2008)*

Children More Susceptible to Air Pollution Than Adults

In many health effects research studies, children are considered as if they were small adults. This is not really true. There are many differences between children and adults in the ways that they respond to air pollution.

Inhale more than adults.

Children take in more air per unit body weight at a given level of exertion than do adults. When a child is exercising at maximum levels, such as during a soccer game or other sports event, they may take in 20 percent to 50 percent more air and thereby more air pollution than would an adult in comparable activity (Michael T. Kleinman.2000).

May do not show symptoms.

Another important difference is that children do not necessarily respond to air pollution in the same way as adults. Adults exposed to low levels of the pollutant ozone will experience symptoms such as coughing, soreness in their chests, sore throats, and sometimes headaches. Children, on the other hand, may not feel the same symptoms, or at least they do not acknowledge them when asked by researchers. It is currently not known if children actually do not feel the symptoms or if they ignore them while preoccupied with play activities(Michael T. Kleinman.2000).

This probably does not mean that children are less sensitive to air pollution than adults. There are several good studies that show children to have losses in lung functions even when they don't cough or feel discomfort. This is important because symptoms are often warning signals and can be used to trigger protective behavior. Children may not perceive these warning signals and might not reduce their activities on smoggy days.

Spend more time outdoor than adults.

Children also spend more time outside than adults. The average adult, except for those who work mostly outdoors, spends most of their time indoors -- at home, work, or even at the gym. Children spend more time outside, and are often outdoors during periods when air pollution is at its highest.

The typical adult spends 85 percent to 95 percent of their time indoors, while children may spend less than 80 percent of their time indoors(Michael T. Kleinman.2000) . Children may also exert themselves harder than adults when playing outside.

At growing stage.

Perhaps the most important difference between adults and children is that children are growing and developing. Along with their increased body size, children's lungs are growing and changing, too.

Exactly what compounds in the ambient air most affect reproductive and children's health, and how these exposures result in restricted fetal growth, early parturition, and development of respiratory diseases remains largely unknown. The study of air pollution's impact on reproductive outcomes is still a developing area of science with many important questions unanswered, but more evidence is emerging that air pollution exposures in pregnancy and early childhood put children at higher risk of adverse health outcomes. Despite the long history of research linking smoking to poorer birth outcomes and the known similarities in components of cigarette smoke and air pollution, the bulk of all air pollution research targeting reproductive health has been conducted only in the past decade. Recently this research has begun to focus on one specific source of modern-day air pollution -- traffic exhaust(Beate Ritz and Michelle Wilhelm . 2008) .

Air Pollution and Infant Mortality

Only in the 1990s, did studies begin to systematically investigate links between air pollution and infant mortality. These studies largely focused on potential mortality impacts of airborne particulate matter small enough to penetrate into the human respiratory tract, referred to as PM10 (particulate matter less than 10 microns in aerodynamic diameter) and more recently have examined PM2.5, even smaller size particles which can penetrate deep into the lung. Most findings from this research indicated infants living in areas with high levels of these types of particulate matter had a greater risk of mortality during the first year of life, particularly from respiratory causes. In a study Beate Ritz and Michelle Wilhelm between 1989 and 2000, found higher risks of infant mortality for very young infants (1-3 months of age) breathing high levels of carbon monoxide (CO), and that infants 4-12 months of age exposed to high levels of particulate matter (PM10) were at higher risk of death from respiratory illnesses. Furthermore, infants exposed to high concentrations of the gaseous pollutant nitrogen dioxide were at increased risk of dying from Sudden Infant Death Syndrome (SIDS)(Beate Ritz and Michelle Wilhelm . 2008) .

Air Pollution and Birth Outcomes

A seminal paper published in 1977 by researchers in Los Angeles was the first to describe a possible association between air pollution at atmospheric levels and reduced birth weight. The studies have linked a number of air pollutants to adverse birth outcomes, including low birth weight (LBW) and small for gestational age (SGA), prematurity, and heart defects at birth. More recently, researchers have also begun to investigate effects on pre-eclampsia (pregnancy-induced hypertension) and spontaneous abortion, the latter possibly triggered by air pollution's damage to DNA in sperm.

In several studies of births during 1990-2003 to women residing in the urban areas, consistently found that mothers exposed to high levels of CO and particles during pregnancy are at higher risk of adverse birth outcomes, including preterm delivery, low birth weight, and congenital heart defects. Researchers found that among all sources of pollutants traffic exhaust be particularly important (Beate Ritz and Michelle Wilhelm . 2008).

Air pollution exposure during early pregnancy may interfere with placental development and subsequent oxygen and nutrient delivery to the fetus throughout pregnancy, while the last trimester is important for fetal weight gain. Exposure to air pollution during specific pregnancy periods may also trigger inflammation and lead to preterm birth (Beate Ritz and Michelle Wilhelm . 2008). Increased risks of certain heart defects were found to be correlated to high levels of carbon monoxide and possibly ozone in the second month of pregnancy, presumably because this is the most important time period for fetal heart development.

To date, scientists have not identified how a mother's exposure to air pollution translates into poorer health of the fetus. Some studies have found that polycyclic aromatic hydrocarbons (PAHs), chemicals formed by the combustion of fossil fuels such as gasoline and diesel, may cause some of the damage. Vehicle exhaust also contains very small or ultra fine particles (UFP), which have a high surface area. PAHs stick to these small particles which are easily inhaled into the human lungs. These UFP can penetrate through the lung barriers into the bloodstream and via the blood, they can enter other organs including the brain and placenta. Studies in mice have shown that particles of the same size as UFP in exhaust can travel to the placenta, chorion, and amniotic fluid, thereby exposing the fetus to these potentially harmful chemicals. These particles can cause inflammation, which may trigger premature labor (Beate Ritz and Michelle Wilhelm. 2008). PAHs may also interfere with placental development and fetal growth early in pregnancy, but more studies are needed to understand the mechanisms that can cause this to happen.

Impact of Air Pollution on Children's Respiratory Health

The lung is an extremely complex organ. While most organs in your body are made up of a few different types of cells, the lung contains more than 40 different kinds of cells. Each of these cells is important to health and maintaining the body's fitness (Michael T. Kleinman.2000).

Air pollution can change the cells in the lung by damaging those that are most susceptible. If the cells that are damaged are important in the development of new functional parts of the lung, then the lung may not achieve its full growth and function as a child matures to adulthood. Although very little research has been conducted to address this extremely important issue, this review will discuss the information that is available.

One of the most comprehensive, long-term studies to date examining the impact of air pollution exposure on children's respiratory health is the Children's Health Study (CHS). Beginning in 1992, University of Southern California researchers collected data from over 6,000 children attending public schools in 12 selected Southern California communities with varying levels of air pollution for a period of 8 or more years. This study has reported on several important findings, including short-term effects of air pollution, such as acute respiratory illnesses and asthma attacks, and longer-term health effects, such as chronic respiratory diseases and development of asthma. For example, researchers found that short-term increases in ozone concentrations were associated with increased school absences due to respiratory illnesses. There is now ample evidence from this and many other studies showing that ozone and particles exacerbate symptoms in asthmatic children. CHS researchers also reported reduced lung function and increased chronic cough and bronchitis in young children and teens chronically exposed to high levels of air pollution, especially those living in areas with high particle concentrations. Southern California children playing many outdoor sports in high-ozone areas were three times more likely to develop asthma. Generally researchers agree that air pollution can trigger asthma attacks and worsen asthma symptoms, but studies have not yet provided unequivocal evidence that it can cause asthma to develop.

Most health effect studies have focused on "criteria" air pollutants regulated under the Clean Air Act, such as CO, ozone, PM10, PM2.5 and NO2, but there is also an increasing awareness that pollutants not routinely measured may be responsible for the detrimental respiratory effects of air pollution, especially UFP from combustion of gasoline and diesel. Animal studies have shown that UFP cause inflammation in respiratory systems and greater allergic reactions, and diesel particles can carry allergens into the body, resulting in a magnified allergic sensitivity and response. Given the evidence these particles may have important effects on the lung, and the fact that children's lungs may be more susceptible to damage, additional research on the effects of ultra fine particles in the context of children's respiratory health are needed(Beate Ritz and Michelle Wilhelm . 2008) .

1. EFFECT OF PARTICULATE MATTER

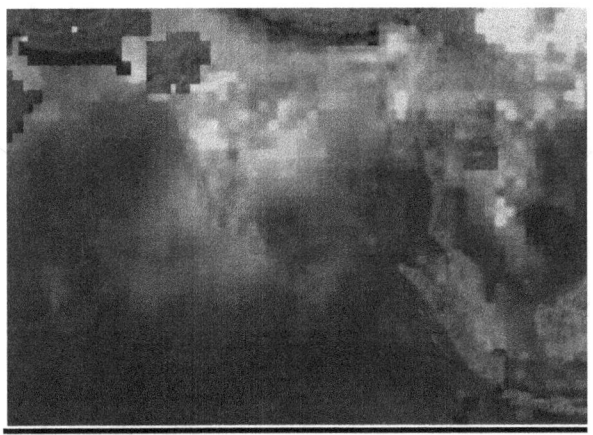

Fig.4.1.1 :Black carbon (soot) aerosol concentration measured during the INDOEX experiment (March 14-21,2001).Yellow =High, Blue = low

Particulate matter (PM) air pollution, measuring less than 2.5 μm, has been the focus of international concern due to its diverse contribution to the global burden of disease. There have been more than 2,000 peer-reviewed studies published since 1997 linking it to strokes, various respiratory and cardiovascular problems and premature death. Unfortunately, the majority of the planet still resides in areas where the World Health Organization Air Quality Guidelines of 10 μg/m3 (annual) and 25 μg/m3 (24-hour period) is exceeded (Brauer M et al.2012).

The size of particles is directly linked to their potential for causing health problems. Small particles less than 10 micrometers in diameter pose the greatest problems, because they can get deep into your lungs, and some may even get into your bloodstream-(Environmental Protection Agency.2008) Even smaller particles, smaller than 2.5 microns in size (PM2.5), are even more likely to lodge and linger in the deepest air sacs of the lung. Penetration of particles is not wholly dependent on their size; shape and chemical composition also play a part. Of course the dangerous needle-like shape of asbestos is widely recognised to lodge itself in the lungs with often dire consequences. Geometrically angular shapes have more surface area than rounder shapes, which in turn affects the binding capacity of the particle to other, possibly more dangerous substances.

Exposure to such particles can affect both lungs and heart. Therefore simple nomenclature can be used to distinguish between the different degrees of relative penetration of a PM particle into the cardiovascular system. Inhalable particles penetrate no further than the bronchi as they are filtered out by the cilia, Thoracic particles can penetrate right into terminal bronchioles whereas PM which can penetrate to alveoli and hence the circulatory system are termed respirable particles. The smallest particles, less than 100 nanometers (nanoparticles), may be even more damaging to the cardiovascular system(Bloomberg L.P. 17 January 2008).

A large number of studies of health impact of suspended particulate air pollution have been undertaken in developing countries(World Health Organization.1997). These studies show remarkable consistency in the relationship observed between changes in daily ambient suspended particulate levels and changes in mortality. Smith(Smith, 1996) estimated the health risk from exposure to particulate air pollution by applying the mean risk per unit ambient concentrations based on the results of some urban epidemiological studies(World Health Organization, Geneva, 1996.)(Hong et al.1997). The range of risk was found to be 1.2 - 4.4% increased mortality per 10 mg/m3 incremental increase in concentration of respirable suspended particles (PM10).

There is evidence that particles smaller than 100 nanometers can pass through cell membranes and migrate into other organs, including the brain. It has been suggested that particulate matter can cause similar brain damage as that found in Alzheimer patients. Particles emitted from modern diesel engines (commonly referred to as Diesel Particulate Matter, or DPM) are typically in the size range of 100 nanometers (0.1 micrometer). In addition, these soot particles also carry carcinogenic components like benzopyrenes adsorbed on their surface. It is becoming increasingly clear that the legislative limits for engines, which are in terms of emitted mass, are not a proper measure of the health hazard. One particle of 10 μm diameter has approximately the same mass as 1 million particles of 100 nm diameter, but it is clearly much less hazardous, as it probably never enters the human body — and if it does, it is quickly removed. Proposals for new regulations exist in some countries, with suggestions to limit the particle surface area or the particle count (numerical quantity).

Increased levels of fine particles in the air as a result of anthropogenic particulate air pollution is consistently and independently related to the most serious effects, including lung cancer (Ole Raaschou-Nielsen et al .July 10, 2013) and other cardiopulmonary mortality. The large number of deaths (National Institute of Environmental Health Sciences) and other health problems associated with particulate pollution was first demonstrated in the early 1970s (Lave, Lester B.; Eugene P. Seskin .1973) and has been reproduced many times since. PM pollution is estimated to cause 22,000-52,000 deaths per year in the United States (from 2000) (Mokdad, Ali H.et al.2004). and contributed to ~370,000 premature deaths in Europe during 2005. (doi:10.2800/165) Numerous scientific studies have linked particle pollution exposure to a variety of problems, including:

- premature death in people with heart or lung disease,
- nonfatal heart attacks,
- irregular heartbeat,
- aggravated asthma,
- decreased lung function, and increased respiratory symptoms, such as irritation of the airways, coughing or difficulty breathing. (www.epa.gov)

People with heart or lung diseases, children and older adults are the most likely to be affected by particle pollution exposure. However, even if you are healthy, you may experience temporary symptoms from exposure to elevated levels of particle pollution. The World Health Organization (WHO) estimates that "... fine particulate air pollution (PM 2.5), causes about 3% of mortality from cardiopulmonary disease, about 5% of mortality from cancer of the trachea, bronchus, and lung, and about 1% of mortality from acute respiratory infections in children under 5 yr, worldwide (doi:10.1080/15287390590936166 PMID 16024504). Researchers suggest that even short-term exposure at elevated concentrations could significantly contribute to heart disease. A study in "The Lancet" concluded that traffic exhaust is the single most serious preventable cause of heart attack in the general public, the cause of 7.4% of all attacks(Nawrot, Tim S; Laura Perez, Nino Künzli, Elke Munters, Benoit Nemery ,2011).

Fig4.1.2: Engulfed PM by alveolar macrophages.(Manas Ranjan Ray and Twisha Lahiri .2010)

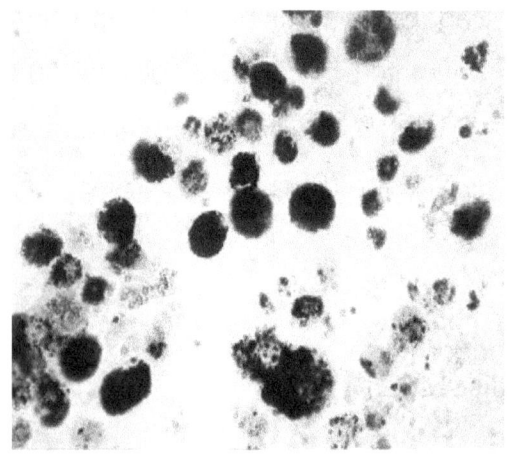 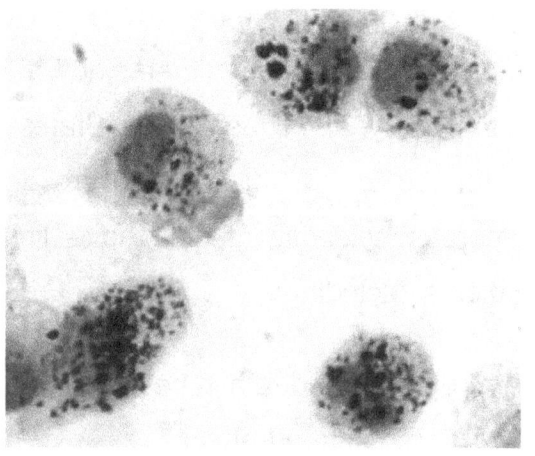

a.Alveolar macrophages engulf inhaled PM carbonaceous PM   *b.Alveolar macrophage with engulfed*

Lung Damage

Great advances have been made in the 1990s in understanding the health effects of fine particles. Since 1987, more than two dozen community health studies have linked respirable particle concentrations below the level of the current air quality standards to reductions in lung function, and increased hospital and emergency room admissions (http://www.nrdc.org/air/transportation/ebd/ebdinx.asp). Long-term exposure has been related to decreases in lung function in both children and adults. Recurrent respiratory illnesses in children are associated with increased particulate exposures, and such a pattern of childhood illness may be a risk factor for later susceptibility to lung damage. Particulate matter exposure causes changes in lung function and inflammation of the small airways. Furthermore, exposure to acidic particles may cause constriction of the bronchi and impair clearance processes which normally remove particles and infectious organisms from the airways. The consequences may include aggravation of existing respiratory problems, more frequent or severe damage to tissues, or greater loss of lung function.

Infections and Asthma

Particulate exposure may increase susceptibility to bacterial or viral respiratory infections, and may increase the incidence of respiratory disease in vulnerable members of the population, including the elderly, people with chronic pulmonary diseases, and people with immune system dysfunction. In the presence of pre-existing heart or lung disease, respiratory exacerbations induced by air pollutants may lead to death. Researches indicates that diesel exhaust may increase the frequency and severity of asthma exacerbations and may lead to inflammation of the airways that can cause or worsen asthma. In cities like Bangalore, around 50% of children suffer from asthma. (The Times of India. 6 November 2007)

Cancer

A review of evidence regarding whether ambient air pollution exposure is a risk factor for cancer in 2007 found solid data to conclude that long-term exposure to PM2.5 (fine particulates) increases the overall risk of nonaccidental mortality by 6% per a 10 microg/m3 increase (PMID 19235364).Exposure to PM2.5 was also associated with an increased risk of mortality from lung cancer (range: 15% to 21% per a 10 microg/m3 increase) and total cardiovascular mortality (range: 12% to 14% per a 10 microg/m3 increase). (PMID 19235364)

Cardiovascular Diseases

According to a recent study by the Registrar General of India and the Indian Council of Medical Research heart disease has emerged as the number one killer among Indians.It is the leading cause of death among males and females in India.India,with more than 1.2 billion people,is estimated to account for 60 percent of heart disease patients wordwide. According to WHO,heart related disorders will kill almost 20 million people by 2015,and these are exceptionally prevalent in the Indian sub-continent.Half of all the heart attacks in this population occur under the age of 50 years and 25 percent under the age of 40.India will have 1.6 million strokes per year by 2015,resulting in disabilities on one third of them (The Tribune,July 31,2013) .With such findings we cannot afford to neglect any single reason behind cardiovascular disease. There are number of studies which shows link between heart maladies and dirty upgrading air pollution.

In the early and middle 1990s researchers began to find, somewhat surprisingly, that air pollution also appeared to be associated with mortality from cardiovascular disease (CVD). Two landmark studies in this area – the Six Cities Study and the American Cancer Society Study – that looked at large populations of people exposed to a wide range of levels of pollution strongly suggested that cardiovascular disease was associated especially with the particulate component (often referred to as "particulate matter" or "PM") of air pollution (Dockery et al. (1993)(Pope, et al. 1995). According to the American Heart Association, 927,448 people died of CVD in the United States in 2002. This represents a death rate of 320.5 per 100,000 people and accounts for 38.0% of all deaths (American Heart Association (2005). Depending on how significant a risk factor air pollution exposure proves to be, relative to the other known risk factors – age, race, lifestyle factors (such as smoking, physical inactivity, and diet), serum lipids, and family history– reduction in air pollution levels could be a means for significantly reducing the burden of CVD. In analysis of the cause of deaths that could be linked to air pollution, it was found that most of the excessive deaths attributed to particulate air pollution were due to heart disease (Dockery. 2001).

In 2004, the first American Heart Association scientific statement on "Air Pollution and Cardiovascular Disease" concluded that exposure to particulate matter (PM) air pollution contributes to cardiovascular morbidity and mortality. The first AHA writing group concluded that short-term exposure to particulate matter (PM) air pollution contributes to acute cardiovascular morbidity and mortality (AHA,2004) and that exposure to elevated PM levels over the long term can reduce life expectancy by a few years. Exposure to PM <2.5 µm in diameter ($PM_{2.5}$) over a few hours to weeks can trigger cardiovascular disease–related mortality and nonfatal events; longer-term exposure (eg, a few years) increases the risk for cardiovascular mortality to an even greater extent than exposures over a few days and reduces life expectancy within more highly exposed segments of the population by several months to a few years; reductions in PM levels are associated with decreases in cardiovascular mortality within a time frame as short as a few years; and many credible pathological mechanisms have been elucidated that lend biological plausibility to these findings. Most, but not all, epidemiological studies corroborate the elevated risk for cardiovascular events associated with exposure to fine PM <2.5 µm in aerodynamic diameter ($PM_{2.5}$). $PM_{2.5}$ generally has been associated with increased risks of myocardial infarction (MI), stroke, arrhythmia, and heart failure exacerbation within hours to days of exposure in susceptible individuals. Several new studies have also demonstrated that residing in locations with higher long-term average PM levels elevates the risk for cardiovascular morbidity and mortality. Some recent evidence also implicates other size fractions, such as ultrafine particles (UFPs) <0.1 µm, gaseous copollutants (eg, ozone and nitrogen oxides [NO_x]), and specific sources of pollution (eg, traffic). In addition, there have been many insights into the mechanisms whereby PM could prove capable of promoting

CVDs. Air pollutants have been linked with endothelial dysfunction and vasoconstriction, increased blood pressure (BP), prothrombotic and coagulant changes, systemic inflammatory and oxidative stress responses, autonomic imbalance and arrhythmias, and the progression of atherosclerosis. In the interim, the US Environmental Protection Agency (EPA) completed its updated "Air Quality Criteria for Particulate Matter" and afterward strengthened the National Ambient Air Quality Standards (NAAQS) for daily $PM_{2.5}$ levels starting in 2006 (down from 65 to 35 µg/m^3). The most recent scientific review coordinated by the EPA, the final report of the Integrated Science Assessment for Particulate Matter (http://cfpub.epa.gov/ncea/cfm/recordisplay.cfm?deid=216546), has also been made available publicly. Researchers at the Johns Hopkins Bloomberg School of Public Health have conducted the largest nationwide study on the acute health effects of coarse particle pollution. The study, published in the May 14, 2008, edition of JAMA, found evidence of an association with hospital admissions for cardiovascular diseases. It has been found that outcome after myocardial infarction show a strong socioeconomic gradient, (Gerber Yet al.2010, Tonne C et al. 2005, Bernheim SM et al 2007, Alter DA et al. 1999, Rao SV t al. 2004, Salomaa V.2000,Chang WC et al.2007, Rasmussen JN et al.2006, Alter D. 2006).although the underlying pathways are not well understood (Lynch JW et al.1996,Diez Roux AV. 2005). Higher levels of air pollution frequently occur in more deprived areas across different populations(O'Neill MS et al. 2003, Molitor J et al .2011 ,Havard S et al. 2009) .This observation raises the possibility that exposure to air pollution may explain, in part, socioeconomic gradients in prognosis among MI patients. In a study with large population of patients with advanced ischaemic heart disease, long-term exposure to air pollution was associated with all-cause mortality and strongest association was found with PM2.5 concentrations (Cathryn Tonne et al.2012).

Proposed Biological Mechanisms

Several mechanisms to explain air pollution toxicity have been proposed and all have data to support them.These ideas developed mostly from research on the effects of PM.

The most obvious explanation is direct toxicity to the cardiovascular tissue. This seems most plausible for

chemicals that can be solubilized in the lung after inhalation, such as aldehydes and metals dissolved from particulate matter, and transported to the cardiovascular tissues, most likely through the blood. However, there is also evidence that the particles can be translocated intact from the lung to vascular and cardiac tissues. This is especially true for the smaller, so-called, ultrafine particles. Once in the vicinity of the CV tissue, the pollutants might adversely affect the tissue by causing or increasing local inflammation, interfering with metabolic processes, disrupting ion channel function, etc. The pollutants might also cause local inflammatory responses in the lung that result in production of inflammatory mediators that could be transported to the cardiovascular tissues, thus affecting function. Either of these two mechanisms

might explain experimental data showing an increase in atherosclerosis in animals exposed to PM (Sun et al .2005).A third hypothesis is that pollutants might trigger nerve endings in the lung which subsequently send impulses to the brain and in turn back to the heart. These nerves are part of the autonomic nervous system and influence how the heart beats.This has been offered as a mechanism to explain evidence that PM in the lung can affect the rhythm of the heart. It is of course likely that all of these mechanisms play a role.At the molecular level, persuasive evidence supports an integral role for ROS-dependent pathways at multiple stages, such as in the instigation of pulmonary oxidative stress, systemic proinflammatory responses, vascular dysfunction, and atherosclerosis(American Heart Association.2010).In sum, new studies continue to support the idea that inhalation of PM can instigate extra pulmonary effects on the cardiovascular system by 3 general "intermediary" pathways. These include pathway 1, the release of proinflammatory mediators (eg, cytokines, activated immune cells, or platelets) or vasculoactive molecules (eg, ET, possibly histamine, or microparticles) from lung-based cells; pathway 2, perturbation of systemic ANS balance or heart rhythm by particle interactions with lung receptors or nerves; and pathway 3, potentially the translocation of PM (ie, UFPs) or particle constituents (organic compounds, metals) into the systemic circulation.

Fig4.1.3: Biological pathways linking PM exposure with CVDs. The 3 generalized intermediary pathways and the subsequent specific biological responses that could be capable of instigating cardiovascular events are shown. MPO indicates myeloperoxidase; PAI, plasminogen activator inhibitor; PSNS, parasympathetic nervous system; SNS, sympathetic nervous system; and WBCs, white blood cells. A question mark (?) indicates a pathway/mechanism with weak or mixed evidence or a mechanism of likely yet primarily theoretical existence based on the literature.(AHA.2010)

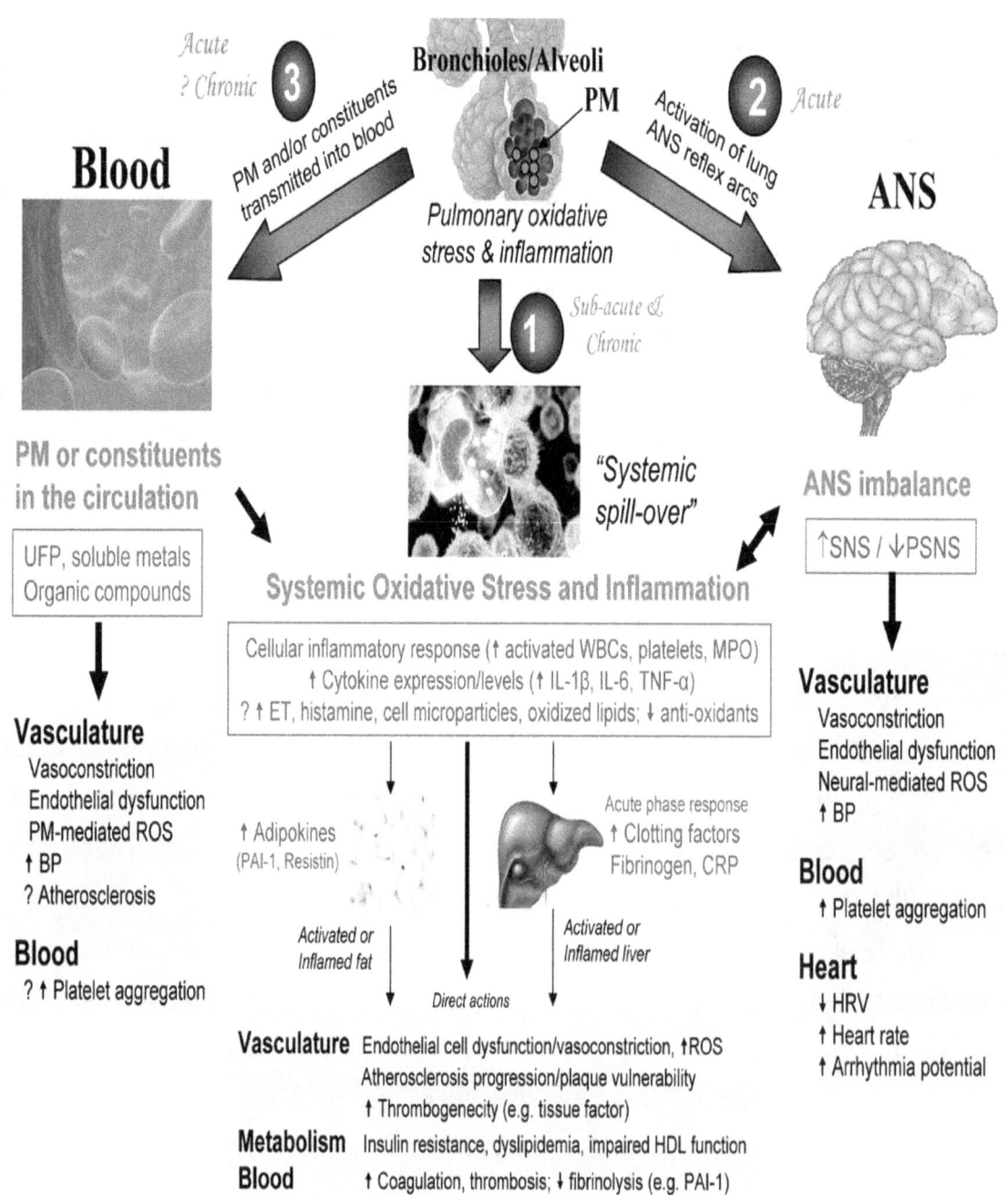

Fine particles (PM$_{2.5}$) cause reduced visibility (haze) (www.epa.gov/visibility).Particles can be carried over long distances by wind and then settle on ground or water. The effects of this settling include: making lakes and streams acidic; changing the nutrient balance in coastal waters and large river basins; depleting the nutrients in soil; damaging sensitive forests and farm crops; and affecting the diversity of ecosystems. Particulate matter can clog stomatal openings of plants and interfere with photosynthesis functions (Hogan, C.Michael .2010). In this manner high particulate matter concentrations in the atmosphere can lead to growth stunting or mortality in some plant species.Particle pollution can also stain and damage stone and other materials, including culturally important objects such as statues and monuments.

Fig4.1.4:Solar radiation reduction due to volcanic eruptions. (wikipedia)

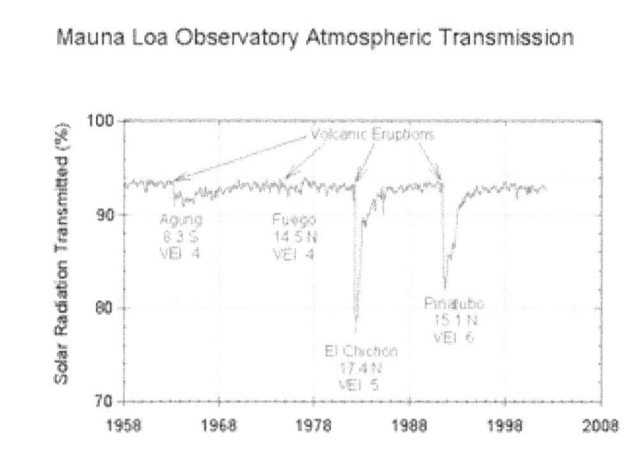

Aerosols are also thought to affect weather and climate on a regional scale. The failure of the Indian Monsoon has been linked to the suppression of evaporation of water from the Indian Ocean due to the semi-direct effect of anthropogenic aerosol (Chung, C E; Ramanathan, V .2006). Recent studies of the Sahel drought (Pollutants and Their Effect on the Water and Radiation Budgets) and major increases since 1967 in rainfall over the Northern Territory, Kimberley, Pilbara and around the Nullarbor Plain have led some scientists to conclude that the aerosol haze over South and East Asia has been steadily shifting tropical rainfall in both hemispheres southward (Chung, C E; Ramanathan, V (2006)(Australian rainfall and Asian aerosols). The latest studies of severe rainfall declines over southern Australia since 1997(Pollution rearranging ocean currents) have led climatologists there to consider the possibility that these Asian aerosols have shifted not only tropical but also midlatitude systems southward.

Aerosol radiative effects

Atmospheric aerosols affect the climate of the earth by changing the amount of incoming solar radiation and outgoing terrestrial long wave radiation retained in the earth's system. This occurs through several distinct mechanisms which are split into direct, indirect(Haywood et al. (2000)(Twomey, S. (1977) and semi-direct aerosol effects. The aerosol climate effects are the biggest source of uncertainty in future climate predictions.(Forster, Piers et al. 2007) The Intergovernmental Panel on Climate Change, Third Assessment Report, says: While the radiative forcing due to greenhouse gases may be determined to a reasonably high degree of accuracy... the uncertainties relating to aerosol radiative forcings remain large, and rely to a large extent on the estimates from global modelling studies that are difficult to verify at the present time.(IPCC Third Assessment Report - Climate Change 2001)

Fig.4.1.4 :Particulates in the air causing shades of grey and pink in Mumbai during sunset.(wikipedia)

The Direct aerosol effect consists of any direct interaction of radiation with atmospheric aerosol, such as absorption or scattering. It affects both short and longwave radiation to produce a net negative radiative forcing. (Charlson, R.J.; S E Schwartz, J M Hales, R D Cess, J A Coakley, J E Hansen, and D J Hofmann .1992). The magnitude of the resultant radiative forcing due to the direct effect of an aerosol is dependent on the albedo of the underlying surface, as this affects the net amount of radiation absorbed or scattered to space. e.g. if a highly scattering aerosol is above a surface of low albedo it has a greater radiative forcing than if it was above a surface of high albedo. The converse is true of absorbing aerosol, with the greatest radiative forcing arising from a highly absorbing aerosol over a surface of high albedo (Haywood et al (2000). The Direct aerosol effect is a first order effect and is therefore classified as a radiative forcing by the IPCC (Forster et al. 2007). The interaction of an aerosol with radiation is quantified by the Single Scattering Albedo (SSA), the ratio of scattering alone to scattering plus absorption (*extinction*) of radiation by a particle.

Fig 4.1.6:An Asian dust cloud during the spring of 2001. The dust cloud was generated by high winds over China's Gobi Desert.

Change to the earth's radiative budget due to the modification of clouds by atmospheric aerosols, and several distinct effects are produced due to PM. Cloud droplets form onto pre-existing aerosol particles, known as cloud condensation nuclei (CCN).For any given meteorological conditions, an increase in CCN leads to an increase in the number of cloud droplets. This leads to more scattering of shortwave radiation i.e. an increase in the albedo of the cloud, known as the Cloud albedo effect, First indirect effect or Twomey effect (Twomey et al (1977). Evidence supporting the cloud albedo effect has been observed from the effects of ship exhaust plumes (Ackerman et al (1995) and biomass burning (Kaufman, Y. J.; Fraser, Robert S. 1997) on cloud albedo compared to ambient clouds. The Cloud albedo aerosol effect is a first order effect and therefore classified as a radiative forcing by the IPCC (Forster et al. 2007). An increase in cloud droplet number due to the introduction of aerosol acts to reduce the cloud droplet size, as the same amount of water is divided between more droplets. This has the effect of suppressing precipitation, increasing the cloud lifetime, known as the cloud lifetime aerosol effect, second indirect effect or Albrecht effect (Forster et al. 2007). This has been observed as the suppression of drizzle in ship exhaust plume compared to ambient clouds (Ferek et al.2000), and inhibited precipitation in biomass burning plumes (Rosenfeld, D .1999). This cloud lifetime effect is classified as a climate feedback (rather than a radiative forcing) by the IPCC due to the interdependence between it and the hydrological cycle (Forster et al. 2007. However, it has previously been classified as a negative radiative forcing (Hansen, J.; Sato, M.; Ruedy, R. 1997).

Figure 4.1.5: Schematic diagram showing the various radiative mechanisms associated with cloud effects that have been identified in relation to aerosols. The small black dots represent aerosol particles, the larger open circles cloud droplets. Straight lines represent the incident and reflected solar radiation, and wavy lines represent terrestrial radiation. The filled white circles indicate cloud droplet number concentration. The unperturbed cloud contains larger cloud drops because only natural aerosols are available as cloud condensation nuclei, while the perturbed cloud contains a greater number of smaller cloud drops because both natural and anthropogenic aerosols are available as cloud condensation nuclei. The vertical grey dashes represent rainfall, and LWC refers to liquid water content. (Forster et al. 2007)

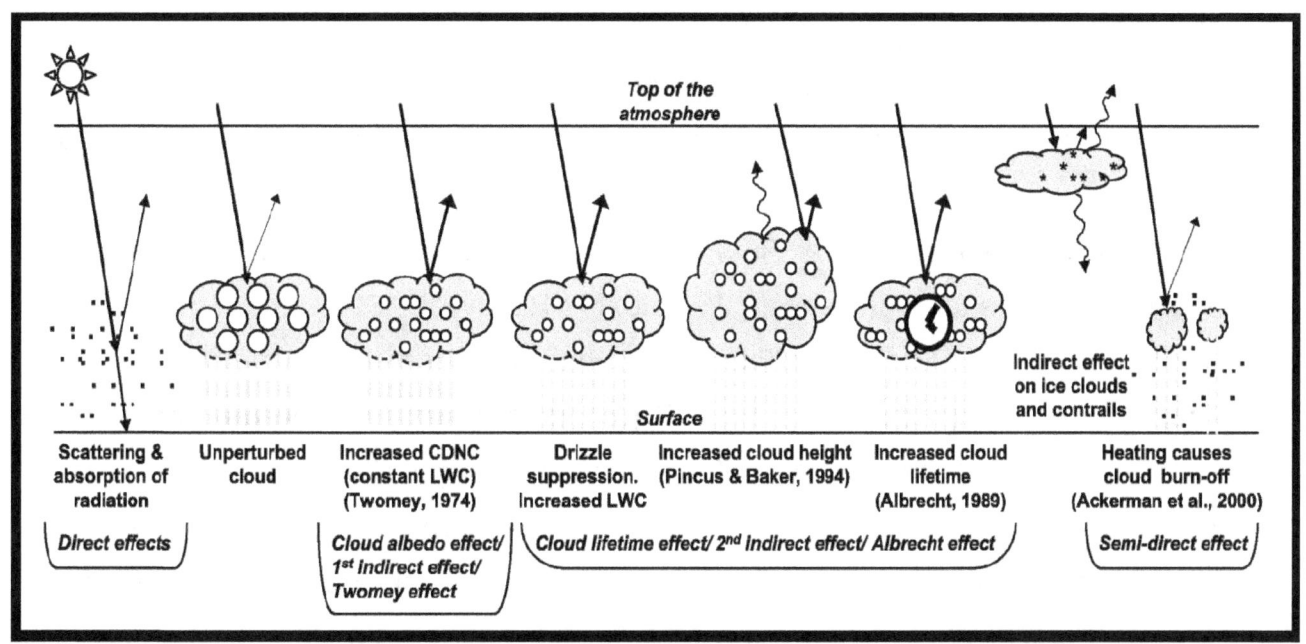

> **"Climate Effects of Black Carbon Aerosols in China and India"**
>
> *(Surabi Menon, James Hansen, Larissa Nazarenko, Yunfeng Luo)*
>
> In recent decades, there has been a tendency toward increased summer floods in south China, increased drought in north China, and moderate cooling in China and India while most of the world has been warming. We used a global climate model to investigate possible aerosol contributions to these trends. We found precipitation and temperature changes in the model that were comparable to those observed if the
> aerosols included a large proportion of absorbing black carbon ("soot"), similar to observed amounts. Absorbing aerosols heat the air, alter regional atmospheric stability and vertical motions, and affect the large scale circulation and hydrologic cycle with significant regional climate effects (Science, 27 September 2002).

There are some other effects too. For instance, if absorbing aerosols are present in a layer aloft in the atmosphere, they can heat surrounding air which inhibits the condensation of water vapour, resulting in less cloud formation. (Ackerman et al. 2000.) Additionally, heating a layer of the atmosphere relative to the surface results in a more stable atmosphere due to the inhibition of atmospheric convection. This inhibits the convective uplift of moisture,(Koren et al. 2004.) which in turn reduces cloud formation.

The heating of the atmosphere aloft also leads to a cooling of the surface, resulting in less evaporation of surface water. The effects described here all lead to a reduction in cloud cover i.e. an increase in planetary albedo. The semi-direct effect classified as a climate feedback by the IPCC due to the interdependence between it and the hydrological cycle(Forster et al. 2007). However, it has previously been classified as a negative radiative forcing(Hansen, J.; Sato, M.; Ruedy, R. 1997). Each type of aerosol has its own effect . Such as,sulfate aerosol has two main effects, direct and indirect. The direct effect, via albedo, is a cooling effect that slows the overall rate of global warming: the IPCC's best estimate of the radiative forcing is -0.4 watts per square meter with a range of -0.2 to -0.8 W/m² (IPPCC. 2001) but there are substantial uncertainties. The effect varies strongly geographically, with most cooling believed to be at and downwind of major industrial centres. Modern climate models addressing the attribution of recent climate change take into account sulfate forcing, which appears to account (at least partly) for the slight drop in global temperature in the middle of the 20th century. The indirect effect (via the aerosol acting as cloud condensation nuclei, CCN, and thereby modifying the cloud properties -albedo and lifetime-) is more uncertain but is believed to be a cooling.. Black carbon from fossil fuels is estimated by the IPCC in the Fourth Assessment Report of the IPCC, 4AR, to contribute a global mean radiative forcing of +0.2 W/m² (was +0.1 W/m² in the Second Assessment Report of the IPCC, SAR), with a range +0.1 to +0.4 W/m².

2. EFFECTS OF SULFUR OXIDES.

Sulfur dioxide is one of the major air pollutants and has significant impacts upon human health (Sulfur Dioxide. United States Environmental Protection Agency). Inhaling sulfur dioxide is associated with increased respiratory symptoms and disease, difficulty in breathing, and premature death(Sulfur Dioxide U.S. Environmental Protection Agency). In 2008, the American Conference of Governmental Industrial Hygienists reduced the short-term exposure limit from 5 ppm to 0.25 ppm. The OSHA PEL is currently set at 5 ppm (13 mg/m³) time weighted average. NIOSH has set the IDLH at 100 ppm(NIOSH

Pocket Guide to Chemical Hazards). A 2011 systematic review concluded that exposure to sulfur dioxide is associated with preterm birth (Shah PS et al 2011). Sulfur trioxide will cause serious burns on both inhalation and ingestion since it is highly corrosive and hygroscopic in nature. SO_3 should be handled with extreme care as it reacts with water violently and produces highly corrosive sulfuric acid.

In addition the concentration of sulfur dioxide in the atmosphere can influence the habitat suitability for plant communities as well as animal life (C.Michael Hogan. 2010). Sulphur dioxide can damage vegetation and cause corrosion. Sulfur dioxide emissions are a precursor to acid rain and atmospheric particulates. Sulfur compounds were responsible for the traditional wintertime sulphur smog in London in the mid 20th century. These anthropogenic pollutants have sometimes reached lethal concentrations in the atmosphere, such as during the infamous London episode of December 195213.

Considering depth of the problem US EPA undertaken Acid Rain Program,for desulfurization in which calcium oxide (lime) reacts with sulfur dioxide to form calcium sulfite ($CaO + SO_2 \rightarrow CaSO_3$).Aerobic oxidation of the $CaSO_3$ gives $CaSO_4$, anhydrite.this desulfurization lead to 33% less emission between 1983 and 2002.Sulfur can be removed from coal during the burning process by using limestone as a bed material in Fluidized bed combustion.(Lindeburg, Michael R.2006).Sulfur can also be removed from fuels prior to burning the fuel, preventing the formation of SO_2 because there is no sulfur in the fuel from which SO_2 can be formed. The Claus process is used in refineries to produce sulfur as a byproduct. The Stretford process has also been used to remove sulfur from fuel. Redox processes using iron oxides can also be used, for example, Lo-Cat(Lo-Cat Process) or Sulferox(SFA Pacific 2002). Fuel additives, such as calcium additives and magnesium oxide, are being used in gasoline and diesel engines in order to lower the emission of sulfur dioxide gases into the atmosphere (Walter R. May Marine Emissions Abatement).

3. IMPACT OF NITROGEN OXIDES

When NO_x and volatile organic compounds (VOCs) react in the presence of sunlight, they form photochemical smog, a significant form of air pollution, especially in the summer. Children, people with lung diseases such as asthma, and people who work or exercise outside are particularly susceptible to adverse effects of smog such as damage to lung tissue and reduction in lung function (United States Environmental Protection Agency). NO_x reacts with ammonia, moisture, and other compounds to form nitric acid vapor and related particles. Small particles can penetrate deeply into sensitive lung tissue and damage it, causing premature death in extreme cases. Inhalation of such particles may cause or worsen respiratory diseases, such as emphysema or bronchitis, or may also aggravate existing heart disease (Environmental protection agency).

Mono-nitrogen oxides are also involved in tropospheric production of ozone (D. Fowler, et al. (1998). NO_x reacts with volatile organic compounds in the presence of sunlight to form Ozone. Ozone can cause adverse effects such as damage to lung tissue and reduction in lung function mostly in susceptible populations (children, elderly, asthmatics). Ozone can be transported by wind currents and cause health impacts far from the original sources. The American Lung Association estimates that nearly 50 percent of United States inhabitants live in counties that are not in ozone compliance (Ozone, Environmental Protection Agency).

NO_x destroys ozone in the stratosphere (NOAA.August 27, 2009) Ozone in the stratosphere absorbs ultraviolet light, which is potentially damaging to life on earth ("Ozone layer". Retrieved Sept,2007). NO_x from combustion sources does not reach the stratosphere. instead, NO_x is formed in the stratosphere from photolysis of nitrous oxide (NOAA, August 27, 2009). NO_x also readily reacts with common organic chemicals, and even ozone, to form a wide variety of toxic products: nitroarenes, nitrosamines and also the nitrate radical some of which may cause biological mutations. Recently another pathway, via NO_x, to ozone has been found that predominantly occurs in coastal areas via formation of nitryl chloride when NO_x comes into contact with salt mist (Carol Potera .2008).

4. ACTION OF CARBON MONOXIDE.

Carbon monoxide poisoning is the most common type of fatal air poisoning in many countries(Omaye ST (2002). Carbon monoxide is colourless, odorless, and tasteless, but highly toxic. It combines with hemoglobin to produce carboxyhemoglobin, which usurps the space in hemoglobin that normally carries oxygen, but is ineffective for delivering oxygen to bodily tissues. Carbon monoxide is absorbed through breathing and enters the blood stream through gas exchange in the lungs. Normal circulating levels in the blood are 0% to 3%, and are higher in smokers. Carbon monoxide levels cannot be assessed through a physical exam. Laboratory testing requires a blood sample (arterial or venous) and laboratory analysis on a CO-Oximeter. Additionally, a noninvasive carboxyhemoglobin (SpCO) test method from Pulse CO-Oximetry exists and has been validated compared to invasive methods (Roth D et al .2011). Concentrations as low as 667 ppm may cause up to 50% of the body's hemoglobin to convert to carboxyhemoglobin (Tikuisis et al. (1992). A level of 50% carboxyhemoglobin may result in seizure, coma, and fatality. In the United States, the OSHA limits long-term workplace exposure levels above 50 ppm (OSHA CO guidlines). Within short time scales, carbon monoxide absorption is cumulative, since the half-life is about 5 h in fresh air.

Carbon monoxide is produced naturally by the human body as a signaling molecule. Thus, carbon monoxide may have a physiological role in the body, such as a neurotransmitter or a blood vessel relaxant (Wu, L; Wang, R .December 2005). Because of carbon monoxide's role in the body, abnormalities in its metabolism have been linked to a variety of diseases, including neurodegenerations, hypertension, heart failure, and inflammation. (Wu, L; Wang, R .December 2005).As Carbon monoxide interferes with blood's ability to carry oxygen. With the blood stream carrying less oxygen, brain function is affected and heart rate increases in an attempt to offset the oxygen deficit. Breathing between 20 to 35 ppm CO in air for 4 h results in impairment of time related response. Individuals with heart condition may experience chest pain.The most common symptoms of carbon monoxide poisoning may resemble other types of poisonings and infections, including symptoms such as headache, nausea, vomiting, dizziness, fatigue, and a feeling of weakness. Affected families often believe they are victims of food poisoning. Infants may be irritable and feed poorly. Neurological signs include confusion, disorientation, visual disturbance, syncope and seizures (Blumenthal, Ivan .1 June 2001). Carbon monoxide also binds to other molecules such as myoglobin and mitochondrial cytochrome oxidase. Exposures to carbon monoxide may cause significant damage to the heart and central nervous system, especially to the globus pallidus, (Prockop LD, Chichkova RI ,2007). often with long-term sequelae. Carbon monoxide may have severe adverse effects on the fetus of a pregnant woman(Tucker Blackburn, Susan .2007).

Some descriptions of carbon monoxide poisoning include retinal hemorrhages, and an abnormal cherry-red blood hue (Ganong, William F.2005). In most clinical diagnoses these signs are seldom noticed (Blumenthal, Ivan .1 June 2001). One difficulty with the usefulness of this cherry-red effect is that it corrects, or masks, what would otherwise be an unhealthy appearance, since the chief effect of removing deoxygenated hemoglobin is to make an asphyxiated person appear more normal, or a dead person appear more lifelike, similar to the effect of red colorants in embalming fluid. The "false" or unphysiologic red-coloring effect in anoxic CO-poisoned tissue is related to the meat-coloring commercial use of carbon monoxide, discussed below.

5. EFFECT OF LEAD

Lead is a highly poisonous metal (regardless if inhaled or swallowed), affecting almost every organ and system in the body. The main target for lead toxicity is the nervous system, both in adults and children. Exposure to Pb can affect the blood, kidneys, and nervous, immune, cardiovascular, and reproductive systems.Measurements made in exposed communities indicate that lead concentration of 1 μg/m3 in ambient air results in an increase of about 1-2 μg per decilitre (μg/dl) in blood. Lead poisoning can cause destructive behavioural changes, learning disabilities and permanent brain damage. Children and pregnant women are at greatest risk. Blood levels of 50 – 60 μg/dl are associated with neurobehavioural changes in children.Long-term exposure of adults can result in decreased performance in some tests that measure functions of the nervous system ("Lead in Air"). Long-term exposure to lead or its salts (especially soluble salts or the strong oxidant PbO_2) can cause nephropathy, and colic-like abdominal pains. It may also cause weakness in fingers, wrists, or ankles. Lead exposure also causes small increases in blood pressure, particularly in middle-aged and older people and can cause anemia. Exposure to high lead levels can severely damage the brain and kidneys in adults or children and ultimately cause death. In pregnant women, high levels of exposure to lead may cause miscarriage. Chronic, high-level exposure have shown to reduce fertility in males (Golub, Mari S., ed. 2005). Lead also damages nervous connections (especially in young children) and cause blood and brain disorders. Lead poisoning typically results from ingestion of food or water contaminated with lead; but may also occur after accidental ingestion of contaminated soil, dust, or lead-based paint (Agency for Toxic Substances and Disease Registry/Division of Toxicology and Environmental Medicine. 2006). It is rapidly absorbed into the bloodstream and is believed to have adverse effects on the central nervous system, the cardiovascular system, kidneys, and the immune system (Bergeson, Lynn L. 2008). The component limit of lead (1.0 µg/g) is a test benchmark for pharmaceuticals, representing the maximum daily intake an individual should have. However, even at this low level, a prolonged intake can be hazardous to human beings (Heavy Metals Testing By Usp. Caspharma.com) (pharmaceutical – Britannica Online Encyclopedia. Britannica.com).

The concern about lead's role in cognitive deficits in children has brought about widespread reduction in its use (lead exposure has been linked to learning disabilities)(Hu, Howard .1991). Most cases of adult elevated blood lead levels are workplace-related. High blood levels are associated with delayed puberty in girls (Schoeters et al .2008). Lead has been shown many times to permanently reduce the cognitive capacity of children at extremely low levels of exposure (Needleman et al .1990). Lead can be found in harmful quantities in stoneware,(Grandjean, P. (1978). vinyl (Levin, R.. et al. (2008) (such as that used for tubing and the insulation of electrical cords), and Chinese brass. White lead paint has been withdrawn from sale in industrialized countries, but the yellow lead chromate is still in use. Old paint should not be stripped by sanding, as this produces inhalable dust.(Marino, P et al .1990). Lead salts used in pottery glazes have on occasion caused poisoning, when acidic drinks, such as fruit juices, have leached lead ions out of the glaze (U.S. Food and Drug Administration. Retrieved feb,2010). It has been suggested that what was known as "Devon colic" arose from the use of lead-lined presses to extract apple juice in the manufacture of cider. Lead is considered to be particularly harmful for women's ability to reproduce. Lead(II) acetate (also known as sugar of lead) was used in the Roman Empire as a sweetener for wine, and some consider this a plausible explanation for the dementia of many Roman emperors, and, that chronic lead poisoning contributed to the empire's gradual decline. (see Decline of the Roman Empire#Lead poisoning)(Angier, Natalie .August 21, 2007).

Exposure to lead and lead chemicals can occur through inhalation, ingestion and dermal contact. Most exposure occurs through ingestion or inhalation. Lead exposure is a global issue as lead mining and lead smelting are common in many countries. Most countries have stopped using lead-containing gasoline by 2007(Agency for Toxic Substances and Disease Registry.2011). Lead can be ingested through fruits and vegetables contaminated by high levels of lead in the soils they were grown in. Soil is contaminated through particulate accumulation from lead in pipes, lead paint and residual emissions from leaded gasoline that was used before the Environment Protection Agency issue the regulation around 1980 . The use of lead for water pipes is problematic in areas with soft or (and) acidic water. Hard water forms insoluble layers in the pipes while soft and acidic water dissolves the lead pipes (Moore, Michael R. 1977). Inhalation is the second major pathway of exposure, especially for workers in lead-related occupations. Almost all inhaled lead is absorbed into the body, the rate is 20–70% for ingested lead; children absorb more than adults(Agency for Toxic Substances and Disease Registry.2011). Dermal exposure may be significant for a narrow category of people working with organic lead compounds, but is of little concern for general population. The rate of skin absorption is also low for inorganic lead (Agency for Toxic Substances and Disease Registry.2011).

Biochemistry of poisoning

In the human body, lead inhibits porphobilinogen synthase and ferrochelatase, preventing both porphobilinogen formation and the incorporation of iron into protoporphyrin IX, the final step in heme synthesis. This causes ineffective heme synthesis and subsequent microcytic anemia(Cohen .1981). At lower levels, it acts as a calcium analog, interfering with ion channels during nerve conduction. This is one of the mechanisms by which it interferes with cognition. Acute lead poisoning is treated using disodium calcium edetate: the calcium chelate of the disodium salt of ethylene-diamine-tetracetic acid (EDTA). This chelating agent has a greater affinity for lead than for calcium and so the lead chelate is formed by exchange. This is then excreted in the urine leaving behind harmless calcium(Laurence, D. R. (1966).). According to the Agency for Toxic Substance and Disease Registry, a small amount of ingested lead (1%) will store itself in bones, and the rest will be excreted by an adult through urine and feces within a few weeks of exposure. However, only about 32% of lead will be excreted by a child.("Toxic Substances Portal – Lead". Agency for Toxic Substance and Disease Registry)

6. BAD FACE OF HYDROCARBONS.

Hydrocarbons (HC) , along with nitrogen oxide, play a significant role as ozone precursors. Other hydrocarbons, e.g. benzol, require particular attention because of their carcinogenic effects (SenStadtUmTech 1998) such as benzo[a]pyrene (B[a]P), cause specific cancers, such as small cell carcinoma. Like many other environmental chemicals, B[a]P, in and of itself, is not chemically reactive, but must be metabolically activated by enzymes in our cells. In the case of B[a]P, a polycyclic aromatic hydrocarbon [PAH], it has been well documented that it is formed during the combustion of tobacco or other organic materials and has the potential to cause cancer in animals and man. An elegant set of experiments performed in the 1970s (Holder et al., 1974; Selkirk et al., 1974; Kapitulnick et al., 1977; Levin et al.,1986; Chang et al., 1987) documented that benzo[a]pyrene requires metabolism in mammalian tissues to form the ultimate carcinogenic agent. This reactive intermediate reacts with specific DNA bases in target genes to form DNA-PAH chemical adducts, thereby causing the critical somatic cell mutations that lead to cancer. Many of these studies were performed in animal models and a series of biological oxidative reactions studied in these models were shown to form the 9,10-epoxy-7,8-dihydrodiol-benzo[a]pyrene, the ultimate carcinogenic form of B[a]P (Holder et al.,1974; Levin et al., 1986).

Living matter is exposed to hydrocarbons in many ways directly or indirectly. Some byproducts,

formed during petroleum refining and processing which are used for the manufacturing of other products are highly toxic. Constantly, these toxic compounds are inadvertently released into the environment and if this effect is connected to the effect of accidental crude oil spills worldwide, then these combined sources of unrestricted hydrocarbons constitute the major cause of environmental pollution. The toxicity of the hydrocarbon molecules chemical and physical nature. Petroleum is toxic and can be lethal depending upon the nature of the petroleum fraction, the way of exposure to it, and the time of exposure. Chemicals and dispersants in crude oil can cause a wide range of health effects in people and wildlife, depending on the level of exposure and susceptibility. The highly toxic chemicals contained in crude oil can damage any organ system in the human body like the nervous system, respiratory system, circulatory system, immune system, reproductive system, sensory system, endocrine system, liver, kidney, etc. and consequently can cause a wide range of diseases and disorders (Costello, 1979).

Individuals more susceptible to harm by the toxic effects of crude oil are as follows.

1. Infants, children, and unborn babies.
2. Pregnant women.
3. People with pre-existing serious health problems.
4. People living in conditions that impose health stress.

In the past, considerable research has been directed toward understanding how toxic chemicals cause cancer. The consensus is that foreign compounds are converted to reactive electrophiles that can chemically react with cellular nucleophiles like protein and nucleic acid to form products known as adducts. Such reactions are believed to lead to toxicity by damaging proteins by forming protein adducts or carcinogenesis by forming DNA adducts that result in somatic cell mutations and cancer. However, the literature offers a second role for toxic chemicals in causing abnormal cell proliferation and inflammation that results in formation of aberrant tissue growth and development, such as atherosclerosis. Chemicals such as the a,b-unsaturated aldehydes or reactive electrophilic intermediates of polycyclic aromatic hydrocarbons may be part of this second mechanism for causing disease in humans (Bhatnagar and Srivastava,1992; Ramos et al., 1996; Moorthy et al., 2002). Aberrant proliferation of vascular smooth muscle cells has long been known to allow these cells to calcify and form plaque in arteries and aorta. This aberrant cell growth, accumulation of lipid and calcification eventually results in closure of vessels leading to and throughout the heart, causing cardiovascular disease and atherosclerosis. Therefore, these exposures to reactive chemical intermediates can not only cause cancer, but also other chronic diseases like atherosclerosis, heart disease and possibly other diseases like diabetes. Aldehydes, R-CH=O, like formaldehyde or acrolein formed by decomposition of volatile organic compounds cause toxic endpoints such as atherosclerosis, dyslipidemia (excess lipid in the bloodstream), and endothelium dysfunction (inadequate endothelial cell function). These lead to diseases like diabetes, obesity, hypertension, etc. The research being pursued at the University of Louisville in focuses on how these disease states arise and how they can be prevented. Their particular interest is to see how oxidative and reductive metabolism may blunt these toxic events, protecting against aberrant cell proliferation. A number of enzymes are involved in these oxidation-reduction processes.Aldose reductase (AR) or the aldo-ketoreductase (AKR) enzyme families chemically reduce aldehydes to alcohols, that tend to be less toxic than the aldehyde compounds. In addition, aldehyde dehydrogenases use NAD(P)+ to oxidize aldehydes to carboxylic acids. Both the alcohols and and carboxylic acids can be conjugated with water-soluble compounds that enhance excretion through the kidney or into the bile. They have noted that the hemoprotein family, cytochrome P450 (also known as CYPP450), can also oxidize and reduce aldehydes to carboxylic acids and alcohols as well (Amunom et al., 2005). This enzyme system has high capacity to remove these foreign compounds and appears to serve as a secondary system to protect the body from aldehydes when the capacity of the reductases has been attained.The damage caused by the toxicity of crude oil to organ systems may be immediate or it may take months or years. Singh et al.(2004)

studied the toxicity of fuels with different chemical composition on CD-1 mice (A swiss mice strain that is used as a general purpose stock and an oncological and pharaceutical research. This is a vigorous outbred stock. These mice are fairly docile and easy to handle). The objective of the study was to establish a correlation between the physico- chemical properties of the fuel and their biologic effects on mice. The results of the study demonstrated that the automobile derived diesel exhaust particles were more toxic than the exhaust generated by forklift engines. It was also found that the diesel exhaust particles contain ten times more extractable organic matter than the standard exhaust material generated by forklift engines. A similar type of study conducted by Kinawy (2009) revealed that the inhalation of leaded or unleaded (containing aromatics and oxygenated compounds) gasoline vapors by rats impaired the levels of monoamine neurotransmitters and other biochemical parameters in different areas of the rats' brains. Likewise, several behavioral changes causing aggression in rats were observed. Menkes and Fawcett (1997) discussed the toxicities of lead and manganese added gasoline and the public health hazards due to aromatic and oxygenated compounds in gasoline. The extent of absorption of petroleum components by inhalation, oral, and dermal routes varies significantly because of the wide range of physicochemical properties of these components.

The incorporation of crude oil into the body may affect the reproductive health of humans and to other lives. Obidike et al. (2007) observed that when the male rats were given an oral crude oil treatment using a drenching tube, degeneration and necrosis of interstitial cell occurred followed by the exudation into the interstices in the testes of rats. The study concluded that exposure of rats to crude oil induces reproductive cytotoxicity confined to the differentiating spermatogonia compartment, likewise it may also harm human reproductive cells. The extent of absorption through the various routes depends on the volatility, solubility, and other properties of the specific component or mixture. The more volatile and soluble the oil fractions (low molecular weight aliphatics and light aromatic compounds) are the faster they can leak into groundwater or vaporize into the air. Therefore, living matter may be easily exposed to these crude oil fractions by breathing the contaminated air and by drinking the contaminated water (Welch et al., 1999). As reported by Knox and Gilman (1997), petroleum derived volatiles are one of the causes for the geographically associated childhood cancers.(http://cdn.intechopen.com/pdfs/37042/InTech-Hydrocarbon_pollution_effects_on_living_organisms_remediation_of_contaminated_environments_and_effects_of_heavy_metals_co_contamination_on_bioremediation.pdf).

Methane is an extremely efficient greenhouse gas which contributes to enhanced global warming. Other hydrocarbon VOCs are also significant greenhouse gases via their role in creating ozone and in prolonging the life of methane in the atmosphere, although the effect varies depending on local air quality. Within the NMVOCs, the aromatic compounds benzene, toluene and xylene are suspected carcinogens and may lead to leukemia through prolonged exposure. 1,3-butadiene is another dangerous compound which is often associated with industrial uses. These are also termed as hazardous air pollutants (HAPS), as they cause or may cause cancer or other serious health effects, such as reproductive effects or birth defects.

On the other hand, the non-volatile heavy fractions of crude oil tend to be absorbed by the soil and persist at the site of release, which may harm living beings by skin contact, by intake of contaminated water or food. The heavy fraction of crude oil consists mainly of napthlene-aromatics and poly-aromatic compounds that are carcinogenic and long exposure to these compounds often leads to tumors, cancer, and failure of the nervous system. The aromatic compounds found in petroleum are an important group of environmental pollutants. These aromatic compounds are introduced into the environment from various sources such as natural oil seeps, refinery waste products and emissions, oil storage wastes, accidental spills from oil tankers, petrochemical industrial effluents and emissions, and coal tar processing wastes, etc. Petroleum hydrocarbons can rapidly migrate from the site of contamination and adversely affect terrestrial and aquatic ecosystems and humans.

Crude oils also contain polar organic compounds that contain N, S, and O atoms in various functional groups. The chemical properties of these NSO compounds, particularly their solubility and toxicity, are of environmental concern. It has been documented, that at the sites where oil spills have occurred a portion of the polar organic compounds present in the oil had partitioned into the groundwater rendering high concentration of total petroleum hydrocarbons in drinkable water sources (Mahatnirunkul et al. 2002, Delin et al., 1998, Oudot, 1990). Therefore, drinkable water sources turn out to be unsafe for the human population as well as for the aquatic animals, aquatic plants, and microbes (Griffin and Calder, 1976). Fortunately a diverse group of microorganisms like bacteria, fungi, and yeast may efficiently breakdown crude oil fractions into nontoxic components. There are a large number of studies available on the microbial degradation of crude oil or hydrocarbons (www.intechopen.com. Shukla Abha and Cameotra Swaranjit Singh. "Hydrocarbon Pollution: Effects on Living Organisms, Remediation of Contaminated Environments, and Effects of Heavy Metals Co-Contamination on Bioremediation").

Table4.8.1 :Health effects of Aldehydes and Ketones .(CPCB,2010)

Sl.No	Name of aldehyde/ketone	Health effects
1	Formaldehyde	Irritation of the eyes, respiratory tract , nausea, headache, tiredness, and thirst.
2	Acrolein, Acetaldehyde and Crotonaldehyde	Irritation of the eyes, skin, and mucous membranes of the upper respiratory tract.
3	Propionaldehyde, n-Butyraldehyde and Isobutyraldehyde	Eye and respiratory tract irritation
4	Other high-molecular- weight aldehydes such as Chloroacetaldehyde, valeraldehyde, Furfural, the Butyl aldehydes , Glyoxal, Malonaldehyde, Benzaldehyde, Synapaldehyde	less toxic than formaldehyde and Acrolein
5	methyl ethyl ketone	affects nervous system, creates headaches, dizziness, fatigue, narcosis (acts like a narcotic), nausea, vomiting, and passing out ,irritates the eyes, nose, skin and throat, contact with the eyes can permanently damage them. Repeated exposure may damage the nervous system and may affect the brain

7. OZONE

Ozone contributes to what we typically experience as "smog" or haze. and can cause eye irritation, aggravation of respiratory diseases, and damage to plants and animals.Ground level ozone- what we breathe- can harm our health. Even relatively low levels of ozone can cause health effects. People with lung disease, children, older adults, and people who are active outdoors may be particularly sensitive to ozone. Children are at greatest risk from exposure to ozone because their lungs are still developing and they are more likely to be active outdoors when ozone levels are high, which increases their exposure. Children are also more likely than adults to have asthma (EPA).

Like oxygen, ozone is soluble in the fluids that line the respiratory tract. Therefore some ozone can penetrate into the gas-exchange, or alveolar, region of the deep lung (Michael T. Kleinman.2000).

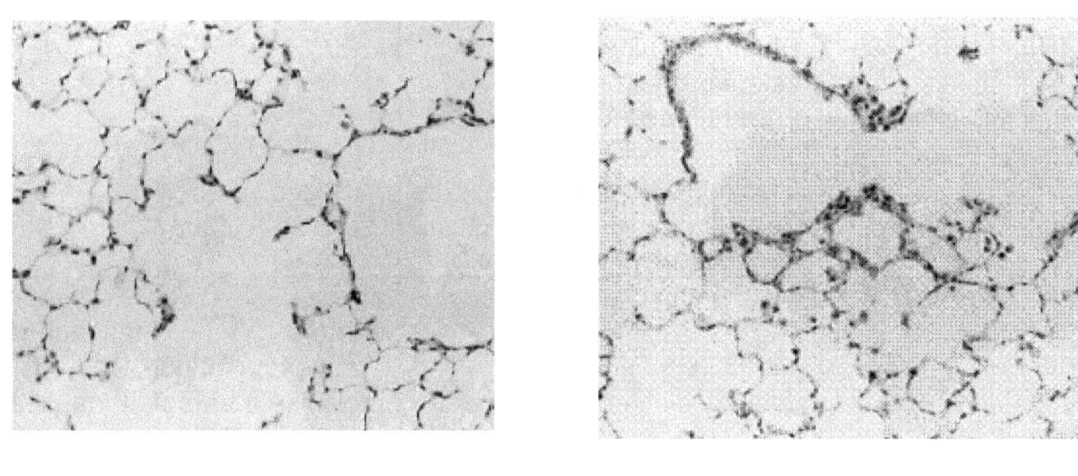

a.   b.

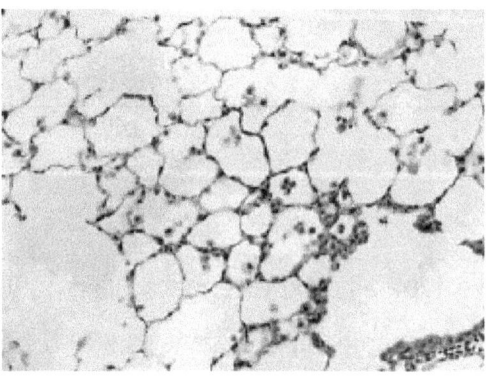

c.

Fig 9.1 : The photos show how ozone affects the sensitive tissue in the deep lung. The pictures are from the lungs of rats exposed to ozone in a laboratory under carefully controlled conditions. The human lung is similar , although not identical to the rat's lung in terms of the types of cells and the overall structure of the alveolar region.(a) Magnified view of the structure of the normal gas-exchange region of the lung.(b) Shows the effects of breathing 0.2 ppm ozone for 4 hours.(c) Shows more extensive damage following exposure to 0.6 ppm of ozone. (Michael T. Kleinman, 2000)

Figure (9.1.a) shows a magnified view of the structure of the normal gas-exchange region of the lung. It is called the gas-exchange region because oxygen inhaled from the air is transferred to the hemoglobin in blood in small blood vessels located inside the thin walls separating the alveolar air spaces. At the same time, carbon dioxide, produced by normal metabolism and dissolved in the blood, is excreted into the air and expired when you breathe out. The walls of a normal alveolus are very thin. There are only two layers of cells and a thin interstitial matrix separating the air in the alveolar space, or lumen, from the fluid inside the blood vessels. The cells that line the healthy alveoli are mostly very broad and very thin, and are called Type I lung cells or Type I pneumocytes. This provides a very large surface area across which gases can be efficiently transported. At the same time, carbon dioxide, produced by normal metabolism and dissolved in the blood, is excreted into the air and expired when you breathe out. The walls of a normal alveolus are very thin. There are only two layers of cells and a thin interstitial matrix separating the air in the alveolar space, or lumen, from the fluid inside the blood vessels. The cells that line the healthy alveoli are mostly very broad and very thin, and are called Type I lung cells or Type I pneumocytes. This provides a very large surface area across which gases can be efficiently transported (Michael T. Kleinman.2000).

Figure (9.1.b) shows the effects of breathing 0.2 ppm ozone for 4 hours. The photo shows evidence of additional cells, called macrophages, and some material that may be fragments of ozone-injured alveolar wall cells inside the alveolar space. Macrophages are immune system cells that respond to the injury of the delicate cells that line the alveolar lumen. These macrophages play important roles in protecting the lungs from inhaled bacteria, fungi and viruses, and are also important in helping to repair lung tissue injury caused by inhaled pollutants (Michael T. Kleinman.2000).

Figure (9.1.c) shows more extensive damage following exposure a higher concentration of ozone, 0.6 ppm. The alveolar walls are thicker and there is evidence of cells infiltrating within the walls. There are more macrophages in the alveolar spaces and the thin, Type I cells have been damaged and replaced with thicker Type II, almost cube-shaped cells that are more resistant to the toxic effects of ozone. All of these changes occurred within 48 hours after exposure. If exposure continues for more than three days, the evidence of cell injury seems to be reduced, except for the continuing presence of the Type II cells (Michael T. Kleinman.2000).

Ozone also affects sensitive vegetation and ecosystems, including forests, parks, wildlife refuges and wilderness areas. In particular, ozone harms sensitive vegetation, including trees and plants during the growing season.

8.HYDROGEN SULFIDE

Hydrogen sulfide is considered a broad-spectrum poison, meaning that it can poison several different systems in the body, although the nervous system is most affected. The toxicity of H_2S is comparable with that of hydrogen cyanide or carbon monoxide(Lindenmann J et al December 2010). It forms a complex bond with iron in the mitochondrial cytochrome enzymes, thus preventing cellular respiration.Since hydrogen sulfide occurs naturally in the body, the environment and the gut, enzymes exist in the body capable of detoxifying it by oxidation to (harmless) sulfate(S. Ramasamy et al .2006). Hence, low levels of hydrogen sulfide may be tolerated indefinitely.At some threshold level, believed to average around 300–350 ppm, the oxidative enzymes become overwhelmed. Many personal safety gas detectors, such as those used by utility, sewage and petrochemical workers, are set to alarm at as low as 5 to 10 ppm and to go into high alarm at 15 ppm.A diagnostic clue of extreme poisoning by H_2S is the discoloration of copper coins in the pockets of the victim.

Exposure to lower concentrations can result in eye irritation, a sore throat and cough, nausea, shortness of breath, and fluid in the lungs (pulmonary edema))(Lindenmann J et al December 2010). These effects are believed to be due to the fact that hydrogen sulfide combines with alkali present in moist surface tissues to form sodium sulfide, a caustic(Lewis, R.J. Sax's Dangerous Properties of Industrial Materials.1996). These symptoms usually go away in a few weeks.

Long-term, low-level exposure may result in fatigue, loss of appetite, headaches, irritability, poor memory, and dizziness. Chronic exposure to low level H_2S (around 2 ppm) has been implicated in increased miscarriage and reproductive health issues among Russian and Finnish wood pulp workers(Hemminki K., Niemi M.L. (1982), but the reports have not (as of circa 1995) been replicated. Although respiratory paralysis may be immediate, it can also be delayed up to 72 hours (http://www.firerescue1.com/fire-products/hazmat-equipment/articles/968922-Thechemical - suicide-phenomenon/).

Different Levels of Hydrogen Sulfide and their Effecst

- 0.00047 ppm or 0.47 ppb is the odor threshold, the point at which 50% of a human panel can detect the presence of the compound(Iowa State University Extension .May 2004).
- 0.0047 ppm is the recognition threshold, the concentration at which 50% of humans can detect the characteristic odor of hydrogen sulfide(Iowa State University Extension .May 2004), normally described as resembling "a rotten egg".
- OSHA has established a permissible exposure limit (PEL) (8 hour time-weighted average (TWA)) of 10 ppm   (Agency for Toxic Substances and Disease Registry .July 2006) .
- 10–20 ppm is the borderline concentration for eye irritation.
- 20 ppm is the acceptable ceiling concentration established by OSHA.
- 50 ppm is the acceptable maximum peak above the ceiling concentration for an 8 hour shift, with a maximum duration of 10 minutes(Agency for Toxic Substances and Disease Registry July 2006). 50–100 ppm leads to eye damage.
- At 100–150 ppm the olfactory nerve is paralyzed after a few inhalations,

and the sense of smell disappears, often together with awareness of danger(USEPA.1980)(Zenz, C., O.B. Dickerson, E.P. Horvath. *Occupational Medicine.* 1994,).

- 320–530 ppm leads to pulmonary edema with the possibility of death (Lindenmann J at al.December 2010).
- 530–1000 ppm causes strong stimulation of the central nervous system and rapid breathing, leading to loss of breathing.
- 800 ppm is the lethal concentration for 50% of humans for 5 minutes exposure (LC50).
- Concentrations over 1000 ppm cause immediate collapse with loss of breathing, even after inhalation of a single breath(Lindenmann J at al.December 2010).

DEADLY H_2S

Hydrogen sulfide was used by the British Army as a chemical weapon during World War I. It was not considered to be an ideal war gas, but, while other gases were in short supply, it was used on two occasions in 1916 (Foulkes, Charles Howard 2001). In 1975, a hydrogen sulfide release from an oil drilling operation in Denver City, Texas killed nine people and caused the state legislature to focus on the deadly hazards of the gas. State Representative E L Short took the lead in endorsing an investigation by the Texas Railroad Commission and urged that residents be warned "by knocking on doors if necessary" of the imminent danger stemming from the gas. One may die from the second inhalation of the gas, and a warning itself may be too late(Howard Swindle, "The Deadly Smell of Success". Retrieved December 14, 2010.). A dump of toxic waste containing hydrogen sulfide is believed to have caused 17 deaths and thousands of illnesses in Abidjan, on the West Africa coast, in the 2006 Côte d'Ivoire toxic waste dump.Three members of one family were killed in a slurry tank in Northern Ireland in September 2012 after one member entered the tank in an attempt to rescue a dog. He was quickly rendered unconscious by the gas (and other toxic gasses), at which point his younger brother entered the tank in an effort to rescue him. He too succumbed to the fumes, and in an attempt to save both sons, the father then entered the tank, and also perished (BBC News. 29 January 2013). The gas, produced by mixing certain household ingredients, was used in a suicide wave in 2008 in Japan (Wired.com .2009).The wave prompted staff at Tokyo's suicide prevention center to set up a special hot line during "Golden Week", as they received an increase in calls from people wanting to kill themselves during the annual May holiday (http://abcnews.go.com/Health/story?id=4908320&page=1).As of 2010, this phenomenon has occurred in a number of US cities, prompting warnings to those arriving at the site of the suicide(http://info.publicintelligence.net/LARTTAChydrogensulfide.pdf).These first responders, such as emergency services workers or family members are at risk of death from inhaling lethal quantities of the gas . (http://www.dhmh.maryland.gov/suicideprevention/ safety%20alert.pdf) (http://www.policemag.com/Channel/Patrol/Articles/Print/Story/2011/04/Duty-DangersChemical -Suicides .aspx) **Local governments have also initiated campaigns to prevent such suicides.**

9. MERCURY

Whether an exposure to the various forms of mercury will harm a person's health depends on a number of factors (below). Almost all people have at least trace amounts of methylmercury in their tissues, reflecting methylmercury's widespread presence in the environment and people's exposure through the consumption of fish and shellfish(http://www.epa.gov/mercury/about.htm). People may be exposed to mercury in any of its forms under different circumstances. The factors that determine how severe the health effects are from mercury exposure include these (http://www.epa.gov/mercury/about.htm):

- the chemical form of mercury;
- the dose;
- the age of the person exposed (the fetus is the most susceptible);
- the duration of exposure;
- the route of exposure -- inhalation, ingestion, dermal contact, etc.; and
- the health of the person exposed.
- Susceptibility factors like immune reactivity,degree of other toxic exposuresand synergism systemic detoxification ability based on blood allele type or metallothionein function,sulfur detoxification deficiencies or other inhinited enzymatic processes related to or methylation (Tingting L et al. 2009).

Mercury exists in three chemical forms. They each have specific effects on human health.

- Methylmercury
- Elemental mercury
- Other mercury compounds (inorganic and organic)

Methylmercury effects

For fetuses, infants, and children, the primary health effect of methylmercury is impaired neurological development. Methylmercury exposure in the womb, which can result from a mother's consumption of fish and shellfish that contain methylmercury, can adversely affect a baby's growing brain and nervous system. Impacts on cognitive thinking, memory, attention, language, and fine motor and visual spatial skills have been seen in children exposed to methylmercury in the womb. Recent human biological monitoring by the Centers for Disease Control and Prevention in 1999 and 2000 (PDF) (3 pp., 42 KB, About PDF) shows that most people have blood mercury levels below a level associated with possible health effects.

Outbreaks of methylmercury poisonings have made it clear that adults, children, and developing fetuses are at risk from ingestion exposure to methylmercury. During these poisoning outbreaks some mothers with no symptoms of nervous system damage gave birth to infants with severe disabilities, it became clear that the developing nervous system of the fetus may be more vulnerable to methylmercury than is the adult nervous system.

In addition to the subtle impairments noted above, symptoms of methylmercury poisoning may include; impairment of the peripheral vision; disturbances in sensations ("pins and needles" feelings, usually in the hands, feet, and around the mouth); lack of coordination of movements; impairment of speech, hearing, walking; and muscle weakness. People concerned about their exposure to methylmercury should consult their physician.

Mercury and Cancer. No human data indicate that exposure to any form of mercury causes cancer, but the human data currently available are very limited. Mercuric chloride has caused increases in several types of tumors in rats and mice, and methylmercury has caused kidney tumors in male mice. Scientists only observed these health effects at extremely high doses, above levels that produced other effects.

Elemental mercury effects

Elemental (metallic) mercury primarily causes health effects when it is breathed as a vapor where it can be absorbed through the lungs. These exposures can occur when elemental mercury is spilled or products that contain elemental mercury break and expose mercury to the air, particularly in warm or poorly-ventilated indoor spaces. Symptoms include these: tremors; emotional changes (e.g., mood swings, irritability, nervousness, excessive shyness); insomnia; neuromuscular changes (such as weakness, muscle atrophy, twitching); headaches; disturbances in sensations; changes in nerve responses; performance deficits on tests of cognitive function. At higher exposures there may be kidney effects, respiratory failure and death. People concerned about their exposure to elemental mercury should consult their physician. (http://www.epa.gov/mercury/about.htm)

Effects of other mercury compounds (inorganic and organic)

High exposures to inorganic mercury may result in damage to the gastrointestinal tract, the nervous system, and the kidneys. Both inorganic and organic mercury compounds are absorbed through the gastrointestinal tract and affect other systems via this route. However, organic mercury compounds are more readily absorbed via ingestion than inorganic mercury compounds. Symptoms of high exposures to inorganic mercury include: skin rashes and dermatitis; mood swings; memory loss; mental disturbances; and muscle weakness. People concerned about their exposure to inorganic mercury should consult their physician (http://www.epa.gov/mercury/effects.htm)

Mercury in the air may settle into water bodies and affect water quality. This airborne mercury can fall to the ground in raindrops, in dust, or simply due to gravity (known as "air deposition"). After the mercury falls, it can end up in streams, lakes, or estuaries, where it can be transferred to methylmercury through microbial activity. Methylmercury accumulates in fish at levels that may harm the fish and the other animals that eat them. Mercury deposition in a given area depends on mercury emitted from local, regional, national, and international sources. The amount of methylmercury in fish in different waterbodies is a function of a number of factors, including the amount of mercury deposited from the atmosphere, local non-air releases of mercury, naturally occurring mercury in soils, the physical, biological, and chemical properties of different waterbodies and the age, size and types of food the fish eats. This explains why fish from lakes with similar local sources of methylmercury can have significantly different methylmercury concentrations.Birds and mammals that eat fish are more exposed to methylmercury than any other animals in water ecosystems. Similarly, predators that eat fish-eating animals are at risk. Methylmercury has been found in eagles, otters, and endangered Florida panthers. Analyses conducted for the Mercury Study Report to Congress suggest that some highly-exposed wildlife species are being harmed by methylmercury. Effects of methylmercury exposure on wildlife can include mortality (death), reduced fertility, slower growth and development and abnormal behavior that affects survival, depending on the level of exposure. In addition, research indicates that the endocrine system of fish, which plays an important role in fish development and reproduction, may be altered by the levels of methylmercury found in the environment.(http://www.epa.gov/mercury/eco.htm)

10. CADMIUM COMPOUNDS

Acute inhalation exposure to high levels of cadmium in humans may result in effects on the lung, such as bronchial and pulmonary irritation. A single acute exposure to high levels of cadmium can result in long-lasting impairment of lung function (ATSDR 1997) (E.J. Calabrese and E.M. Kenyon. 1991) (U.S. Department of Health and Human Services .1993). Cadmium is considered to have high acute toxicity, based on short-term animal tests in rats (U.S. Department of Health and Human Services .1993) .

Chronic inhalation and oral exposure of humans to cadmium results in a build-up of cadmium in the kidneys that can cause kidney disease, including proteinuria, a decrease in glomerular filtration rate, and an increased frequency of kidney stone formation (ATSDR 1997)(E.J. Calabrese and E.M. Kenyon. 1991)(U.S. Department of Health and Human Services .1993). Other effects noted in occupational settings from chronic exposure of humans to cadmium in air are effects on the lung, including bronchiolitis and emphysema (ATSDR 1997)(E.J. Calabrese and E.M. Kenyon. 1991)(U.S. Department of Health and Human Services .1993) . Chronic inhalation or oral exposure of animals to cadmium results in effects on the kidney, liver, lung, bone, immune system, blood, and nervous system (ATSDR 1997)(E.J. Calabrese and E.M. Kenyon. 1991). The Reference Dose (RfD) for cadmium in drinking water is 0.0005 milligrams per kilogram per day (mg/kg/d) and the RfD for dietary exposure to cadmium is 0.001 mg/kg/d; both are based on significant proteinuria in humans. The RfD is an estimate (with uncertainty spanning perhaps an order of magnitude) of a daily oral exposure to the human population (including sensitive subgroups) that is likely to be without appreciable risk of deleterious noncancer effects during a lifetime. It is not a direct estimator of risk, but rather a reference point to gauge the potential effects. At exposures increasingly greater than the RfD, the potential for adverse health effects increases. Lifetime exposure above the RfD does not imply that an adverse health effect would necessarily occur (EPA.1999). EPA has high confidence in both RfDs based primarily on a strong database for cadmium toxicity in humans and animals that also permits calculation of pharmacokinetic parameters of cadmium absorption, distribution, metabolism, and elimination (EPA.1999).

There is some evidence to suggest that maternal cadmium exposure may result in decreased birthweights (ATSDR 1997). Animal studies provide evidence that cadmium has developmental effects, such as low fetal weight, skeletal malformations, interference with fetal metabolism, and impaired neurological development, via inhalation and oral exposure(ATSDR 1997)(E.J. Calabrese and E.M. Kenyon. 1991)(U.S. Department of Health and Human Services .1993) . Limited animal data are available, although some reproductive effects, such as decreased reproduction and testicular damage, have been noted following oral exposures. (ATSDR 1997).

Several occupational studies have reported an excess risk of lung cancer in humans from exposure to inhaled cadmium. Animal studies have reported cancer resulting from inhalation exposure to several forms of cadmium, while animal ingestion studies have not demonstrated cancer resulting from exposure to cadmium compounds(ATSDR 1997)(E.J. Calabrese and E.M. Kenyon. 1991) (EPA 1999). EPA considers cadmium to be a probable human carcinogen (cancer-causing agent) and has classified it as a Group B1 carcinogen(EPA 1999). (http://www.epa.gov/ttn/atw/hlthef/cadmium.html)

11. CHROMIUM

Chromium (VI) is much more toxic than chromium (III), for both acute and chronic exposures (ATSDR.1998) (U.S. EPA 1998) .The respiratory tract is the major target organ for chromium (VI) following inhalation exposure in humans. Shortness of breath, coughing, and wheezing were reported in cases where an individual inhaled very high concentrations of chromium trioxide (ATSDR.1998) (U.S. EPA 1998). Other effects noted from acute inhalation exposure to very high concentrations of chromium (VI) include gastrointestinal and neurological effects, while dermal exposure causes skin burns in humans (ATSDR.1998) (U.S. EPA 1998)(WHO.1988) . Ingestion of high amounts of chromium (VI) causes gastrointestinal effects in humans and animals, including abdominal pain, vomiting, and hemorrhage (ATSDR.1998) . Acute animal tests have shown chromium (VI) to have extreme toxicity from inhalation and oral exposure (ATSDR.1998) (U.S. Department of Health and Human Services .1998) . Chronic inhalation exposure to chromium (VI) in humans results in effects on the respiratory tract, with perforations and ulcerations of the septum, bronchitis, decreased pulmonary function, pneumonia, asthma, and nasal itching and soreness reported (ATSDR.1998) (U.S. EPA 1998)(WHO.1988). Chronic human exposure to high levels of chromium (VI) by inhalation or oral exposure may produce effects on the liver, kidney, gastrointestinal and immune systems, and possibly the blood (ATSDR.1998) (U.S. EPA 1998)(WHO.1988) . Rat studies have shown that, following inhalation exposure, the lung and kidney have the highest tissue levels of chromium (ATSDR.1998) (U.S. EPA 1998)(WHO.1988). Dermal exposure to chromium (VI) may cause contact dermatitis, sensitivity, and ulceration of the skin (ATSDR.1998) (U.S. EPA 1998)(WHO.1988). The Reference Concentration (RfC) for chromium (VI) (particulates) is 0.0001 mg/m^3 based on respiratory effects in rats. The RfC is an estimate (with uncertainty spanning perhaps an order of magnitude) of a continuous inhalation exposure to the human population (including sensitive subgroups) that is likely to be without appreciable risk of deleterious noncancer effects during a lifetime. It is not a direct estimator of risk but rather a reference point to gauge the potential effects. At exposures increasingly greater than the RfC, the potential for adverse health effects increases (U.S. EPA.1999) . The Reference Dose (RfD) for chromium (VI) is 0.003 mg/kg/d based on the exposure at which no effects were noted in rats exposed to chromium in the drinking water. EPA has low confidence in the RfD based on: low confidence in the study on which the RfD for chromium (VI) was based because a small number of animals were tested, a small number of parameters were measured, and no toxic effects were noted at the highest dose tested; and low confidence in the database because the supporting studies are of equally low quality and developmental endpoints are not well studied (U.S. EPA.1999) .Chromium (III) is an essential element in humans, with a daily intake of 50 to 200 µg/d recommended for adults (ATSDR.1998) .Acute animal tests have shown chromium (III) to have moderate toxicity from oral exposure (ATSDR.1998))(U.S. Department of Health and Human Services .1998). Although data from animal studies have identified the respiratory tract as the major target organ for chronic chromium exposure, these data do not demonstrate that the effects observed following inhalation of chromium (VI) particulates are relevant to inhalation of chromium (III) (U.S. EPA.1999).

Epidemiological studies of workers have clearly established that inhaled chromium is a human carcinogen, resulting in an increased risk of lung cancer. Although chromium-exposed workers were exposed to both chromium (III) and chromium (VI) compounds, only chromium (VI) has been found to be carcinogenic in animal studies, so it has been concluded that only chromium (VI) should be classified as a human carcinogen (ATSDR.1998) (U.S. EPA.1999) . Animal studies have shown chromium (VI) to cause lung tumors via inhalation exposure (ATSDR.1998) (WHO.1998). EPA has classified chromium (VI) as a Group A, known human carcinogen by the inhalation route of exposure (U.S. EPA.1999). EPA has stated that "the classification of chromium (VI) as a known human carcinogen raises a concern for the carcinogenic potential of chromium (III)" (U.S. EPA.1999). (http://www.epa.gov/ttn/atw/hlthef/chromium.html)

12. ASBESTOS

Chronic inhalation exposure to asbestos in humans can lead to a lung disease called asbestosis, which is a diffuse fibrous scarring of the lungs. Symptoms of asbestosis include shortness of breath, difficulty in breathing, and coughing. Asbestosis is a progressive disease, i.e., the severity of symptoms tends to increase with time, even after the exposure has stopped. In severe cases, this disease can lead to death, due to impairment of respiratory function (ATSDR.1995) (E.J. Calabrese and E.M. Kenyon. 1991).Other effects from asbestos exposure via inhalation in humans include pulmonary hypertension and immunological effects (ATSDR.1995) (E.J. Calabrese and E.M. Kenyon. 1991).

A large number of occupational studies have reported that exposure to asbestos via inhalation can cause lung cancer and mesothelioma (a rare cancer of the membranes lining the abdominal cavity and surrounding internal organs) (ATSDR.1995) (E.J. Calabrese and E.M. Kenyon. 1991) (U.S. Department of Health and Human Services .1993). Individuals who smoke and are also exposed to asbestos have a greater than additive increased risk of developing lung cancer.Several occupational studies have reported an increase in gastrointestinal cancer from inhalation exposure to asbestos and subsequent oral ingestion (ATSDR.1995) (E.J. Calabrese and E.M. Kenyon. 1991) . Long- and intermediate-range asbestos fibers (>5 micrometers (μm)) appear to be more carcinogenic than short fibers (<5 μm) (ATSDR.1995) . Several epidemiological studies have found an association between asbestos in drinking water and cancer of the esophagus, stomach, and intestines; however confounding factors and the short follow up time relative to the long latent period for tumor formation make it difficult to interpret the results (ATSDR.1995) (U.S. EPA.1999) .A series of large-scale lifetime feeding studies in animals reported that intermediate-range asbestos fibers increased the incidence of a benign tumor of the large intestine in male rats, while short-range asbestos fibers showed no significant increase in tumor incidence(ATSDR.1995) (U.S. EPA.1999) . EPA considers asbestos to be a human carcinogen (cancer-causing agent) and has ranked it in EPA's Group A (U.S. EPA.1999) (http://www.epa.gov/ttn/atw/hlthef/asbestos.html).

13.EFFECTS OF DIOXINS AND FURANS.

Gruesome Colour of Death

Agent Orange is the code name for one of the herbicides and defoliants used by the U.S. military as part of its herbicidal warfare program, Operation Ranch Hand, during the Vietnam War from 1961 to 1971. Vietnam estimates 400,000 people being killed or maimed, and 500,000 children born with birth defects.(York,Geoffrey; Mick, Hayley.2008)

A 50:50 mixture of 2,4,5-T and 2,4-D, it was manufactured for the U.S. Department of Defenseprimarily by Monsanto Corporation and Dow Chemical. The 2,4,5-T used to produce Agent Orange was later discovered to be contaminated with 2,3,7,8-tetrachlorodibenzodioxin, an extremely toxic dioxin compound. It was given its name from the color of the orange-striped 55 US gallon (200 L) barrels in which it was shipped, and was by far the most widely used of the so-called "Rainbow Herbicides"(Hay, 1982). During the Vietnam War, between 1962 and 1971, the United States military sprayed nearly 20,000,000 US gallons (75,700,000 L) of chemical herbicides and defoliants in Vietnam, eastern Laos and parts of Cambodia, as part of Operation Ranch Hand.[3][4] The program's goal was to defoliate forested and rural land, depriving guerrillas of cover; another goal was to induce forced draft urbanization, destroying the ability of peasants to support themselves in the countryside, and forcing them to flee to the U.S. dominated cities, thus depriving the guerrillas of their rural support base and food supply(Pellow, David N.2007,Stellman et al. 2003,Kolko, Gabriel .1994).

The US began to target food crops in October 1962, primarily using Agent Blue. In 1965, 42 percent of all herbicide spraying was dedicated to food crops(Kolko, Gabriel (1994). Rural-to-urban migration rates dramatically increased in South Vietnam, as peasants escaped the destruction and famine in the countryside by fleeing to the U.S.-dominated cities. The urban population in South Vietnam nearly tripled: from 2.8 million people in 1958, to 8 million by 1971. The rapid flow of people led to a fast-paced and uncontrolled urbanization; an estimated 1.5 million people were living in Saigon slums, while many South Vietnamese elites and U.S. personnel lived in luxury(Luong, 2003).

United States Air Force records show that at least 6,542 spraying missions took place over the course of Operation Ranch Hand(Furukawa, Hisao (2004). By 1971, 12 percent of the total area of South Vietnam had been sprayed with defoliating chemicals, at an average concentration of 13 times the recommended USDA application rate for domestic use. In South Vietnam alone, an estimated 10 million hectares of agricultural land were ultimately destroyed(Luong, 2003:). In some areas TCDD concentrations in soil and water were hundreds of times greater than the levels considered "safe" by the U.S. Environmental Protection Agency(Fawthrop, Tom.2004,2008) Overall, more than 20% of South Vietnam's forests were sprayed at least once over a nine year period (Kolko, Gabriel .1994).

Effects on the Vietnamese people

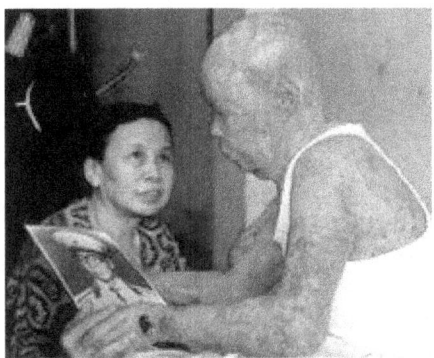

(Above .Major Tự Đức Phang was exposed to dioxin-contaminated Agent Orange)

The Vietnam Red Cross reported as many as 3 million Vietnamese people have been affected by Agent Orange, including at least 150,000 children born with birth defects(Department of Veterans Affairs Office of Public Health and Environmental Hazards. March 25, 2010. Retrieved May 4, 2010.). According to Vietnamese Ministry of Foreign Affairs, 4.8 million Vietnamese people were exposed to Agent Orange, resulting in 400,000 people being killed or maimed, and 500,000 children born with birth defects(York,Geoffrey; Mick, Hayley.2008). Children in the areas where Agent Orange was used have been affected and have multiple health problems, including cleft palate, mental disabilities, hernias, and extra fingers and toes(*BBC News*. December 3, 1998). In the 1970s, high levels of dioxin were found in the breast milk of South Vietnamese women, and in the blood of U.S. soldiers who had served in Vietnam(Thornton, Joe .2001). . The most affected zones are the mountainous area along Truong Son (Long Mountains) and the border between Vietnam and Cambodia. The affected residents are living in substandard conditions with many genetic diseases.(Vietnam Ministry of Foreign Affairs – *Support Agent Orange Victims* in Vietnamese.

)

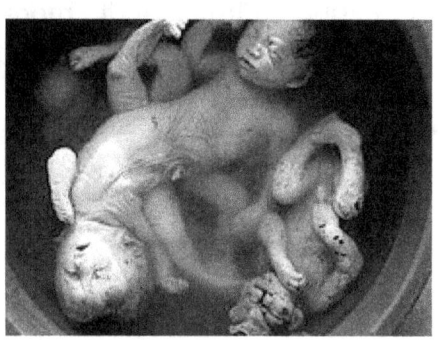

Vietnamese babies, deformed andstillborn after prenatal dioxin exposure from Agent Orange

About 28 of the former US military bases in Vietnam where the herbicides were stored and loaded onto airplanes may still have high level of dioxins in the soil, posing a health threat to the surrounding communities. Extensive testing for dioxin contamination has been conducted at the former US airbases in Da Nang, Phu Cat and Bien Hoa. Some of the soil and sediment on the bases have extremely high levels of dioxin requiring remediation. The Da Nang Airbase has dioxin contamination up to 350 times higher than international recommendations for action(Hatfield Consultants.2007,2009) The contaminated soil and sediment continue to affect the citizens of Vietnam, poisoning their food chain and causing illnesses, serious skin diseases and a variety of cancers in the lungs, larynx, and prostate(*BBC News*. December 3, 1998).

Ecological effects

About 17.8% (3,100,000 ha) of the total forested area of Vietnam was sprayed during the war, which dramatically disrupted ecological equilibrium. Furthermore, the persistent nature of dioxins, erosion caused by loss of protective tree cover, and loss of seeding forest stock, meant reforestation was difficult or impossible in many areas(Furukawa, 2004). Many defoliated forest areas were quickly invaded by aggressive pioneer species, such as bamboo and cogon grass, which make it unlikely the forests will be able to regenerate. Animal species diversity was also significantly impacted: in one study, a Harvard biologist found 24 species of birds and 5 species of mammals in a sprayed forest, while in two adjacent sections of unsprayed forest there were 145 and 170 species of birds and 30 and 55 species of mammals(Chiras, Daniel D. (2010).). Dioxins from Agent Orange have persisted in the Vietnamese environment since the war, settling in the soil and sediment and entering into food chain through the animals and fish that feed in the contaminated areas. Movement of dioxins through the food web has resulted inbioconcentration and biomagnification(Vallero, Daniel A. (2007).). The areas most heavily contaminated with dioxins are the sites of former U.S. air bases(Furukawa, 2004).

Sociopolitical effects

The RAND Corporation's *Memorandum 5446-ISA/ARPA* states: "the fact that the VC obtain most of their food from the neutral rural population dictates the destruction of civilian crops ... if they (the VC) are to be hampered by the crop destruction program, it will be necessary to destroy large portions of the rural economy – probably 50% or more".

Rural-to-urban migration rates dramatically increased in South Vietnam, as peasants escaped the destruction in the countryside by fleeing to the U.S.-dominated cities. The urban population in South Vietnam more than tripled: from 2.8 million people in 1958, to 8 million by 1971. The rapid flow of people led to a fast-paced and uncontrolled urbanization; an estimated 1.5 million people were living in Saigon slums, while many South Vietnamese elites and U.S. personnel lived in luxury(Luong, 2003).

Effects on U.S. veterans

Studies showed that veterans who served in the South during the war have increased rates of cancer, and nerve, digestive, skin and respiratory disorders. Veterans from the south had higher rates of throat cancer, acute/chronic leukemia, Hodgkin's lymphoma and non-Hodgkin's lymphoma, prostate cancer, lung cancer, soft tissue sarcoma and liver cancer. Other than liver cancer, these are the same conditions the US Veteran's Administration has found to be associated with exposure to Agent Orange/dioxin, and are on the list of conditions eligible for compensation and treatment(Department of Veterans Affairs Office of Public Health and Environmental Hazards. March 25, 2010.

Retrieved May 4, 2010). Military personnel who loaded airplanes and helicopters used in Ranch Hand probably sustained some of the heaviest exposures. Members of the Army Chemical Corps, who stored and mixed herbicides and defoliated the perimeters of military bases, are also thought to have had some of the heaviest exposures. Others with potentially heavy exposures included members of U.S. Army Special Forces units who defoliated remote campsites, and members of U.S. Navy river units who cleared base perimeters (Frumkin, 2003). Military members who served on Okinawa also claim to have been exposed to the chemical(Mitchell, Jon.2011). While in Vietnam, the veterans were told not to worry, and were persuaded the chemical was harmless(Hermann, Kenneth J.2006). After returning home, Vietnam veterans began to suspect their ill health or the instances of their wives having miscarriages or children born with birth defects might be related to Agent Orange and the other toxic herbicides to which they were exposed in Vietnam. Veterans began to file claims in 1977 to the Department of Veterans Affairs for disability payments for health care for conditions they believed were associated with exposure to Agent Orange, or more specifically, dioxin, but their claims were denied unless they could prove the condition began when they were in the service or within one year of their discharge.By April 1993, the Department of Veterans Affairs had only compensated 486 victims, although it had received disability claims from 39,419 soldiers who had been exposed to Agent Orange while serving in Vietnam (Fleischer, Doris Zames; Zames, Freida .2001)

Dioxins and furans can cause a number of health effects. The most well known member of the dioxins/furans family is 2,3,7,8 TCDD. The U.S. Environmental Protection Agency (EPA) has said that it is likely to be a cancer causing substance to humans. In addition, people exposed to dioxins and furans have experienced changes in hormone levels (http://www.epa.gov/osw/hazard/wastemin/minimize/factshts/dioxfura.pdf).After systematically evaluating the epidemiologic studies and rodent bioassays on TCDD, U.S.EPA report utilized a TCDD physiologically-based pharmacokinetic model to simulate TCDD blood concentrations, the dose metric used in the dose-response analyses. The draft report develops an oral reference dose (RfD) of 7×10^{-10} mg/kg-day based on two epidemiologic studies that associated TCDD exposures with decreased sperm concentration and sperm motility in men who were exposed during childhood (Mocarelli et al., 2008, 199595) and increased thyroid-stimulating hormone levels in newborn infants (Baccarelli et al., 2008, 197059). High doses of dioxin have caused a skin disease called chloracne.

Dioxin exposure has been linked to birth defects, inability to maintain pregnancy, decreased fertility, reduced sperm counts, endometriosis, diabetes, learning disabilities, immune system suppression, lung problems, skin disorders, lowered testosterone levels and much more (http://www.ejnet.org/dioxin/)It was observed that more girls than boys are born in some Canadian communities because airborne pollutants called dioxins can alter normal sex ratios, even if the source of the pollution is many kilometers away, researchers say. Dioxin exposure has been shown elsewhere to lead to both higher cancer rates and the birth of more females. Researchers at the IntrAmericas Centre for Environment and Health say their findings, confirm the phenomenon in Canada (Oct 18,2007).The study also reveals the health risks of living within 25 km (15.5 miles) of sources of pollution -- a greater distance than previously thought, they said. Normally, 51 percent of births are boys and 49 percent are girls. But the ratio was reversed -- with as few as 46 males born for every 54 females -- in Canadian cities and towns where parents were exposed to pollutants from sources such as oil refineries, paper mills and metal smelters, according to the study. "If you find an inverted sex ratio, and want to know what causes it, look for sources of dioxin," said James Argo, a medical geographer who headed the study, which was published in a journal of the American Chemical Society. "In every one of those cities where those industries are found, there was a higher probability of female births to male births," Argo said in an interview. Using birth data and an inventory of pollution sources, the study also concluded that early exposure to dioxins ,even at 25 km away from the source ,increased the risk of cancer later in life in a group of 20,000 people surveyed during the 1990s. Previous studies that linked dioxins with cancer and a gender imbalance focused on smaller distances, usually about 5 km, Argo said. Dioxins are toxic chemicals found in very small amounts in the air, water, soil and some foods. The large-scale burning of municipal and medical waste is the primary source of dioxins in Canada, but they are also created by fuel and wood burning, electrical power generation, and in the production of iron and steel. Since more females were born in the 90 communities studied, more breast, uterine, cervical and ovarian cancers were observed among them than other forms of cancer, Argo said.

The EPA report confirmed that dioxin is a cancer hazard to people. In 1997, the International Agency for Research on Cancer (IARC) -- part of the World Health Organization -- published their research into dioxins and furans and announced on February 14, 1997, that the most potent dioxin, 2,3,7,8-TCDD, is a now considered a Group 1 carcinogen, meaning that it's a known human carcinogen. Also, in January 2001, the U.S. National Toxicology Program upgraded 2,3,7,8-TCDD from "Reasonably Anticipated to be a Human Carcinogen" to "Known to be a Human Carcinogen." (11th Report on Carcinogens ,related documents under 2,3,7,8-Tetrachlorodibenzo-p-dioxin (TCDD) and Furan)(http://www.ejnet.org/dioxin/). A July 2002 study shows dioxin to be related to increased incidence of breast cancer. Finally, a 2003 re-analysis of the cancer risk from dioxin reaffirmed that there is no known "safe dose" or "threshold" below which dioxin will not cause cancer. (http://www.ejnet.org/dioxin/) EPA classifies TCDD as carcinogenic to humans, based on numerous lines of evidence, including primarily: multiple occupationally- and accidentally-exposed epidemiologic cohorts showing an association between TCDD exposure and certain cancers or increased mortality from all cancers and extensive evidence of carcinogenicity at multiple tumor sites in both sexes of multiple species of experimental animals. (U.S. EPA.2010).

DIRTY DIESEL EXHAUST AND CANCER

WHO, International Agency for Research on Cancer, U.S. EPA, Cal EPA, the National Institute of Occupational Safety and Health and CSE India have all consistently agreed on the relationship between diesel exhaust exposure and lung cancer. Numerous independent analyses of the data by top scientists have come to the same conclusions. From the scientific evidences it is clear: diesel exhaust is a complex mixture comprised of hazardous particles and vapors, some of which are known carcinogens and others probable carcinogens. The IARC has reclassified diesel exhaust and removed it from Group 2A list of `probable carcinogens` to its Group 1 list of substances that have definite links to cancer – thus changing its status to 'carcinogen'. Diesel exhaust is now in the same class of deadly carcinogens as asbestos, arsenic or tobacco among others. 2 The World Health Organization's recent Global Burden of Disease (GBD) Assessment estimates that outdoor air pollution causes 620,000 premature deaths per year in India, a six fold increase since 2000. The main causes are growing emissions of particulate emissions (PM10) from transport and power plants. GBD in this analysis has ranked air pollution as the sixth most dangerous killer in South Asia and fifth leading cause of deaths in India. Cancers as a group account for approximately 13% of all deaths each year with the most common being: lung cancer (1.3 million deaths). Besides this, problem is exaggerated as diesel consumption is increasing year per year in our country.According to CSE there is huge leap in availability of diesel variant vehicles for last few years. Dieselization has taken off at a maniacal pace with state subsidies. Despite recession, diesel cars have clocked 34 per cent growth last year and are close to 60 per cent of new car sales. On the other hand ,the IARC-WHO has urged worldwide efforts to reduce exposure to diesel fumes as much as possible..Also, according to the WHO, across the G-20 economies, 13 of the 20 most polluted cities are in India and over 50% of the sites studied across India had critical levels of PM10 pollution. These findings are indeed worrisome. Average annual concentration in Delhi for example is about 120 µg / m3, as against a residential National Standard of 60 µg / m3 and WHO guideline of 20 µg / m3 (*www.nrdc.org*). More concerning fact is that according to study conducted by Dr.Dale Hattis ,Clark University among a million people exposed chronically to 1 microgram per cubic meter ($\mu g/m^3$) of diesel exhaust, Dr. Hattis's estimated 90 percent confidence range indicates that 34 to 650 people might be expected to develop lung cancer. This risk is far above U.S. EPA's "negligible risk" level as well as deadly when exposure is 60 times more to most of the urban residentials in our country. In a city like Delhi, more than 55 per cent of its 17 million people live within 500 metres of major roads and are directly affected by traffic emissions. India's cancer registry says cancer is taking on an epidemic form that demands immediate action to cut environmental risks. A World Bank supported study on source apportionment of PM2.5 in selected Indian cities, released in 2004, shows that depending on the season, the contribution of diesel fuel to the total PM2.5 ambient concentration can be as high as 61 per cent in Kolkata, 23 per cent in Delhi and 25 per cent in Mumbai. The cancer-causing potential of diesel exhaust is several times higher than some of the worst known air toxics. For instance, the number of excess cancer cases per million people per microgram per cubic meter diesel particulate emissions concentration over a 70-year lifetime exposure is 300. This is several times higher than dangerous toxics like 1,3-Butadiene which is 170 (*Cseindia.org.2012*). Smoking has been banned in public places but are we aware that diesel exhaust shares carcinogens with smoking fumes (Table I).Certainly emission by bulky diesel engines is many folds higher and inhaled everyday by almost every one of us. For eg benzene, arsenic and beryllium compounds. According to IARC Diesel exhaust is potential enough for various forms of DNA damage, including bulky adducts, oxidative damage, strand breaks, unscheduled synthesis, mutations, sister chromatid exchange, morphological cell transformation in mammalian cells, and mutations in bacteria. Increased expression of genes involved in xenobiotic metabolism, oxidative stress, inflammation, antioxidant response, apoptosis, and cell cycle regulation in mammalian cells is also observed. Positive genotoxicity biomarkers of exposure and effect is another observation in humans exposed to diesel engine exhaust. All these findings are certainly a wake-up call to India which is in the grip of rapid dieselization, under-priced and under-taxed toxic diesel and facing failure of government agencies and the state owned enterprises in discharging their Constitutional and Statutory duties. Not only cancer Available evidence indicates that current exposure levels are high enough to lead to other adverse health effects. Diesel exhaust can affect the immune system ,respiratory system, produce allergic reactions. Some reports of individuals in the workplace and in clinical studies exposed acutely to high concentrations of diesel exhaust have shown neurophysiological symptoms such as headache, light-headedness, nausea, vomiting, and numbness or tingling of the extremities(WHO,CSE, NIOSH). There has been some evidence from animal studies indicating possible neurological and behavioural effects. There have also been a few studies in animals showing sperm abnormalities, neurobehavioural effects in pups and other effects on reproduction as well as on sex ratio.

Table I. Detailed comparison of carcinogens present in cigarette and diesel exhaust.

Cigarette Carcinogens	Diesel exhaust Carcinogens	
Arsenic	Contaminant	Note
Benzene	acetaldehyde	IARC Group 2B carcinogens
Beryllium (a toxic metal)	acrolein	IARC Group 3 carcinogens
1,3–Butadiene (a hazardous gas)	aniline	IARC Group 3 carcinogens
Cadmium	antimony compounds	Toxicity similar to arsenic poisoning
Chromium (a metallic element)	arsenic	IARC Group 1 Carcinogens, endocrine disruptor
Ethylene oxide	benzene	IARC Group 1 Carcinogens
Nickel (a metallic element)	beryllium compounds	IARC Group 1 Carcinogens
Polonium-210 (a radioactive chemical element)	biphenyl	It has mild toxicity.
Vinyl chloride	bis(2-ethylhexyl)phthalate	endocrine disruptor
	1,3-butadiene	IARC Group 2A carcinogens
Other toxic chemicals in secondhand smoke are suspected to cause cancer, including (1):	cadmium	IARC Group 1 Carcinogens, endocrine disruptor
	chlorine	
Formaldehyde	chlorobenzene	It has "low to moderate" toxicity.

Benzo[α]pyrene	chromium compounds	IARC Group 3 carcinogens
Toluene	cobalt compounds	
	cresol isomers	
	cyanide compounds	
	dibutyl phthalate	endocrine disruptor
	1,8-dinitropyrene	Carcinogen
	dioxins and dibenzofurans	
	ethyl benzene	
	formaldehyde	IARC Group 1 Carcinogens
	inorganic lead	endocrine disruptor
	manganese compounds	
	mercury compounds	IARC Group 3 carcinogens
	methanol	It may cause blindness.
	methyl ethyl ketone	It may cause birth defect.
	naphthalene	IARC Group 2B carcinogens
	nickel	IARC Group 2B carcinogens
	3-nitrobenzanthrone	One of the strongest carcinogens known
	4-nitrobiphenyl	
	phenol	endocrine disruptor
	phosphorus	
	polycyclic organic matter, including polycyclic aromatic hydrocarbons(PAHs)	
	propionaldehyde	
	selenium compounds	IARC Group 3 carcinogens
	styrene	IARC Group 2B carcinogens

	toluene	IARC Group 3 carcinogens
	xylene isomers and mixtures: o-xylenes, m-xylenes, p-xylenes	IARC Group 3 carcinogens

Genotoxic effects and cancer

Diesel particulate matter and extracts of its organic components have induced gene mutations and chromosomal changes in a variety of bacterial and mammalian cell test systems. Both the particle core and the associated organic compounds have demonstrated carcinogenic properties. Many studies in both humans and animals have shown the potential for diesel exhaust to cause or contribute to the development of cancer in the lung. Increased lung cancer risk has been observed in railroad workers, truck drivers, heavy equipment operators, and professional drivers. Several well-conducted studies in the rat have demonstrated that chronic inhalation exposure produced dose-related increases in lung tumours (benign and malignant). However, in other species the evidence is less clear. The consistent findings of carcinogenic activity by the organic extracts of diesel particle matter in non-inhalation studies (intratracheal instillation, lung implantation and skin painting) further contribute to the overall animal evidence. It is biologically plausible for the mutagenic and carcinogenic components of diesel exhaust to increase the risk of lung cancer. This supports a causal relation between the association observed between exposures and cancers. The International

Agency for Research on Cancer (IARC, 1989) which classified diesel exhaust in group 2A – probably carcinogenic to humans has now convinced with the fact that it is carcinogen and therefore reclassified it in group 1(IARC,2012).

Mechanisms of carcinogenicity : According to IARC Diesel exhaust is potential enough for various forms of DNA damage, including

- bulky adducts,

- oxidative damage,

- strand breaks,

- unscheduled synthesis,

- mutations,

- sister chromatid exchange,

- morphological cell transformation in mammalian cells, and mutations in bacteria.

Increased expression of genes involved in xenobiotic metabolism, oxidative stress, inflammation, antioxidant response, apoptosis, and cell cycle regulation in mammalian cells was observed.

- Positive genotoxicity biomarkers of exposure and effect were also observed in humans exposed to diesel engine exhaust. (IARC)

Other Non-Cancer Impacts

Many of the individual constituents of diesel exhaust are known to produce harmful effects. Benzene, for example, is known to cause disorders of the blood and the blood-forming tissues. Formaldehyde and acetaldehyde can cause irritation of the eyes, nose, and throat. Toluene, lead, cadmium, and mercury are known to cause birth defects and other reproductive problems. Dioxins are toxic to the immune system, interfere with hormone function, and are toxic to reproduction. These non-cancer effects of diesel exhaust components can also be serious and damaging. The extent to which these effects may occur from current exposure levels is unclear.Diesel exhaust is of low acute toxicity, however exposure can result in death from carbon monoxide, a component of diesel exhaust (http://www.toronto.ca/health/pdf/de_technical_ appendix.pdf). Acute exposure to diesel exhaust has been associated with irritation of the eye, nose, and throat, and with respiratory symptoms such as cough and phlegm. Diesel exhaust is known to contain various irritants in both the gaseous phase and particulate phase (for example, sulphur oxides (SOx), nitrogen oxides (NOx) and aldehydes). The evidence for potential chronic noncancerhealth effects of diesel exhaust is based primarily on findings from chronic animal inhalation studies showing chronic inflammation and tissue changes in the lung in rats, mice, hamsters and monkeys(Ishinishi, N. et al., 1986.) . A few studies of workers have noted some respiratory symptoms, but overall, available studies have not shown significant chronic non-cancer health effects associated with diesel exhaust exposure in humans. Several epidemiological studies have demonstrated an association between air pollution and day-to-day changes in mortality, hospital emergency visits, and changes in lung function. (Heinrich, U. et al., 1995. Ishinishi, N. et al., 1988.)Some studies in animals have shown decreased immune function after exposure to diesel exhaust, but others have not. Recent human and animal studies have shown that short exposures to diesel exhaust can produce allergic reactions and exacerbate symptoms to otherallergens (toxprobe).Some reports of individuals in the workplace and in clinical studies exposed acutely to high concentrations of diesel exhaust have shown neurophysiological symptoms such as headache, light-headedness, nausea, vomiting, and numbness or tingling of the extremities. There has been some evidence from animal studies indicating possible neurological and behavioural effects. However, these have been observed at exposures higher than those have that caused respiratory effects (http://www.toronto.ca/health/pdf/de_technical_appendix.pdf).

Environmental Fate

The effects of diesel exhaust in the environment are similar to the effects of emissions from burning other fossil fuels. Diesel exhaust contributes to acid deposition (acid rain), the formation of ground-level ozone and global warming. Knowledge concerning the products of chemical transformation of diesel exhaust in the air is still limited. Secondary aerosols such as nitoarenes, nitrates and sulphates from diesel exhaust may also exhibit different biological reactivity than the primary particles. There is evidence that reaction of PAH in the exhaust with nitrogen oxides will form nitroarenes that are often more mutagenic than their precursors. A study has suggested that reaction with ground-level ozone increases the inflammatory effect of diesel particles in the lung of the rat. After being emitted, diesel particles undergo ageing (oxidation, nitration or other chemical and physical changes) in the atmosphere. The atmospheric lifetime of the various compounds found in diesel exhaust varies and ranges from hours to days. Particles that are smaller than 1micron can remain in the atmosphere for up to 15 days. Primary diesel emissions are a complex mixture containing hundreds of organic and inorganic constituents in the gas and particle phases. The more reactive compounds with short atmospheric lifetimes will undergo rapid transformation in the presence of the appropriate reactants, whereas more stable pollutants can be transported over greater distances. The particulate portion of diesel exhaust can be either primary (emitted directly) or secondary (formed from the transformation of the gaseous component). There is little or no hygroscopic growth of primary diesel particles, however products of oxidation are more hygroscopic. Since the products of oxidation are more soluble they are more readily removed from the air. Particles are removed from the atmosphere through accretion of the particles and dry or wet deposition. Particles of small diameter (<1 μm) are removed less efficiently and thus have longer atmospheric residence times. Because of their small size, diesel exhaust particles have residence times in air of several days, and they may be transported over long distances. Ultimately, they may be removed by wet deposition if they serve as condensation nuclei for water vapour deposition or are scavenged by precipitation in or below cloud. Atmospheric lifetimes for several gas-phase components of diesel exhaust are on the order of hours or days, during which time atmospheric turbulence and advection can disperse these pollutants widely. Inorganic species such as sulphur dioxide (SO_2) and nitric acid have relatively fast deposition rates and remain in the air for less time than the organic components. Dry deposition of organic species is typically on the order of weeks to months. Gaseous diesel exhaust will primarily react with sunlight, the hydroxyl (OH) radical, ozone, the hydroperoxyl (HO_2) radical, various nitrogen oxides and sulphuric acid. Reaction with the OH radical is the major removal route for PAHs in the gas phase and occurs within a few hours in daylight. In the presence of nitrogen oxides, this oxidation reaction can lead to the formation of nitroarenes or nitro-PAHs. Oxides of nitrogen (primarily NO) that are emitted in diesel exhaust are also oxidized in the atmosphere to form nitrogen dioxide (NO_2) and particulate nitrate. About 98% of sulphur emitted from diesel engines is in the form of SO_2. This is readily oxidized by the OH radical in the atmosphere and then rapidly transformed into sulphuric acid aerosols (H_2SO_4) through the reaction of the HO_2 radical and HSO_3 with water. Because SO_2 is soluble in water, it is scavenged by fog, cloud water and raindrops. The particle matter of diesel exhaust is primarily composed of carbonaceous material (organic and elemental carbon) with a very small fraction composed of inorganic compounds and metals. The elemental carbon component of diesel exhaust is inert to atmospheric degradation. High-molecular-weight PAHs in particulate matter are generally more resistant to atmospheric reactions than PAHs in the gas phase, leading to an anticipated half-life of 1 or more days. PAHs undergo photolysis, nitration, and oxidation. They react with sunlight, ozone, hydroxyl radicals, nitrogen oxides, nitrates and sulphates. Ultra-fine particles emitted by diesel engines undergo nucleation, coagulation and condensation to form fine particles (http://www.toronto.ca/health/pdf/de_technical_appendix.pdf).

Evidences of carcinogenicity.

Diesel exhaust is one of the most toxic pollutants we breathe every day. It poses 7X the lung cancer risk of *all* other air toxics tracked by EPA combined. It Includes 40 hazardous air pollutants listed under the Clean Air Act, of which 15 are known, probable, or possible carcinogens.

A. IARC-WHO 2012.Organistaion has said that their decision is unanimous and is based on compelling scientific evidence. The most clinching evidence has come from one of the largest American studies in March 2012 year by the US National Cancer Institute. This study has analysed 12,300 miners for several decades starting 1947 and found that miners heavily exposed to diesel exhaust had a higher risk of dying from lung cancer.

B. Attfield et al 2012, Silverman et al 2012. Cohort and nested case–control analyses adjusted for tobacco smoking showed positive trends in lung cancer risk with increasing exposure to diesel exhaust, with 2–3-fold increased risk in the highest categories of cumulative or average exposure.

C.Olsson et al, 2011. Results showed a smoking-adjusted increased risk for lung cancer after exposure to diesel engine exhaust and a positive exposure-response for both a cumulative exposure index and duration of exposure

*D.*ScienceDaily .Sep. 2, 2009 . Scientists have demonstrated that the link between diesel fume exposure and cancer lies in the ability of diesel exhaust to induce the growth of new blood vessels that serve as a food supply for solid tumors. The researchers found that in both healthy and diseased animals, more new blood vessels sprouted in mice exposed to diesel exhaust than did in mice exposed to clean, filtered air. This suggests that previous illness isn't required to make humans susceptible to the damaging effects of the diesel exhaust.

The tiny size of inhaled diesel particles, most less than 0.1 microns in diameter, potentially enables them to penetrate the human circulatory system, organs and tissues, meaning they can do this damage just about anywhere in the body. A micron is one millionth of a meter.

Diesel exhaust exposure levels in the study were designed to mimic the exposure people might experience while living in urban areas and commuting in heavy traffic. The levels were lower than or similar to those typically experienced by workers who use diesel-powered equipment, who tend to work in mines, on bridges and tunnels, along railroads, at loading docks, on farms and in vehicle maintenance garages, according to the U.S. Department of Labor.

"The message from our study is that exposure to diesel exhaust for just a short time period of two months could give even normal tissue the potential to develop a tumor," said Qinghua Sun, senior author of the study and an assistant professor of environmental health sciences at Ohio State University.

"We need to raise public awareness so people give more thought to how they drive and how they live so they can pursue ways to protect themselves and improve their health. And we still have a lot of work to do to improve diesel engines so they generate fewer particles and exhaust that can be released into the ambient air."

The researchers experimented with mice that resembled two conditions that could be present in a human body. In one, the scientists implanted a small platform seeded with normal endothelial cells, the cells that line blood vessels, under the skin of the mice. This was designed to mimic relatively normal conditions in human bodies for cell growth.

In the other, the researchers created an environment that would follow a significant loss of blood flow to a section of a vessel, called ischemia, in the hind limbs of the mice. This generated severe hypoxia, an area with low or no oxygen, a condition that is present in certain diseases.

Both types of mice were then exposed to either whole diesel exhaust containing particles at a concentration of about 1 milligram per cubic meter, or to filtered outdoor air, for six hours per day five days a week. The rest of the time they breathed filtered air in their cages. Effects of the exposure were measured after two weeks, five weeks and eight weeks of the exposures.

Though some blood vessel growth and chemical changes could be seen in the mice after two weeks of exposure, "generally, the longer the exposure, the more effects we could see," said Sun, also an investigator in Ohio State's Davis Heart and Lung Research Institute. "It's difficult to translate outcomes from an animal study directly to the human experience, but the bottom line is, the shorter the exposure to diesel exhaust, the better."

The exposure to diesel exhaust caused a six-fold increase in new blood vessel formation in the ischemic hind limbs after eight weeks and a four-fold increase in vessel sprouting in the normal hind limbs of the mice in the same amount of time, compared to mice breathing filtered air.

The researchers also saw significantly more blood vessel growth in the implanted cells and in rings of tissue taken from the aortas of mice exposed to the exhaust compared to the control mice exposed to clean air. In fact, the researchers found that three types of blood vessel development occurred in these areas after exposure to the diesel exhaust: angiogenesis, the development of new capillaries; arteriogenesis, the maturation or re-started growth of existing vessels; and vasculogenesis, the formation of new blood vessels.

All of these processes are associated with tumor growth, but unprogrammed angiogenesis in particular can wreak havoc in the human body, Sun said.

"Whenever you talk about a solid tumor, angiogenesis is one of the fundamental mechanisms behind its development. Angiogenesis provides the means for tumor cells to grow because they have to have a blood supply. Without a blood supply, solid tumors will not grow," he said.

"We want our bodies to generate new blood vessels only when we need them. And then stop producing them when we need them to stop."

Though the researchers have not defined every mechanism behind these processes, they sought to explain at least a few ways in which blood vessels are able to sprout or mature after exposure to diesel exhaust.

They observed that diesel exhaust exposure activated a chemical signal, vascular endothelial growth factor, which has long been associated with new blood vessel development. The exposure also increased levels of a protein, hypoxia-inducible factor 1, that is essential to blood vessel development when oxygen levels are low. At the same time, the presence of the exhaust lowered the activity of an enzyme that has a role in producing substances that can suppress tumor growth.

The scientists also tracked low-grade inflammation in tissues exposed to the exhaust, which is often associated with tumor development.

Though the tiny size of diesel exhaust particles may contribute to their ability to penetrate all areas of the body, Sun noted that their complex chemical composition, and the way in which those chemicals are released once particles enter the body, also influence how they react with human cells.

Gasoline exhaust particles are larger than diesel fume particles, but it's premature to suggest that they are any less dangerous to humans, Sun said.

"The bigger particles are known to be harmful primarily for upper respiratory tract illnesses. Larger particles also can't travel long distances – they tend to fall to the ground," he said. "Smaller particles hover in the air for a long time and can have long-term impact on humans when inhaled."

E. Mutagenesis. November 2005, pp 399-410. "3-Nitrobenzanthrone, a potential human cancer hazard in diesel exhaust and urban air pollution." Epidemiological studies have shown that exposure to diesel exhaust and urban air pollution is associated with an increased risk of lung cancer. 3-Nitrobenzanthrone [3-nitro-7*H*-benz[*de*]anthracen-7-one (3-NBA)] is an extremely potent mutagen and suspected human carcinogen identified in diesel exhaust and ambient air particulate matter. There is clear evidence that 3-NBA is a genotoxic mutagen forming DNA adducts after metabolic activation through simple reduction of the nitro group. It is concluded that these adducts not only represent premutagenic lesions in DNA but are of primary importance for the initiation of the carcinogenic process and subsequent tumour formation in target tissue. Because of its widespread environmental presence, 3-NBA may represent not only an occupational health hazard but also a hazard for larger sections of the general population. For an accurate risk assessment more epidemiological studies on 3-NBA-exposed individuals and a broader monitoring of environmental levels of 3-NBA are required.

F. Richard B. Schlesinger, Department of Environmental Medicine, New York University Medical Center, New York, USA .Toxicological Evidence for Health Effects from Inhaled Particulate Pollution: Does it Support the Human Experience? Inhalation Toxicology 1995, Vol. 7, No. 1 , pp 99-109. The fine size mode of ambient particulates, designated as PM-10, is a heterogeneous mixture that can vary in particle size and chemical composition, depending on geographical location, meteorology, and source emissions. While epidemiological studies indicate an association between ambient particulate air pollution exposure and increased human mortality and morbidity, the question remains as to biological plausibility. Individual components of PM-10 have been shown, in experimental studies, to produce what may be considered to be adverse health effects similar to those observed in humans. For example, diesel exhaust particles are associated with lung tumors.

G. U.S National Institute of Occupational Safety and Health (niosh) recommends that whole diesel exhaust be regarded as "a potential occupational carcinogen," as defined in the Cancer Policy of the Occupational Safety and Health Administration (OSHA) ("Identification, Classification, and Regulation of Potential Occupational Carcinogens," 29 CFR 1990). This recommendation is based on findings of carcinogenic and tumorigenic responses in rats and mice exposed to whole diesel exhaust.

H. Dale Hattis. 1995. Among a million people exposed chronically to 1 microgram per cubic meter (μg/m^3) of diesel exhaust, Dr. Hattis's estimated 90 percent confidence range indicates that 34 to 650 people might be expected to develop lung cancer. The average estimate is 230 per million so exposed .Unfortunately, most people are exposed to more than 1 μg/m^3 of diesel exhaust every day. Estimates of diesel exhaust exposure levels in urban areas range as high as 23 µg/m³. Chronic exposure at these levels would potentially result in many more lung cancer cases. The U.S. EPA suggests that a cancer risk may be "negligible" if a substance induces one excess cancer out of a million people exposed over a lifetime. Using the mean value in Dr. Hattis's uncertainty distribution for diesel exhaust potency, the expectation is that exposure to the average levels of diesel exhaust found in California-of 1.54 µ/m³ of diesel exhaust-is likely to result in an excess risk over a person's lifetime of about 350 cancers per million exposed. This risk is far above U.S. EPA's "negligible risk" level. Applying these risk estimates, over a lifetime, exposure to diesel exhaust may cause 12,000 *or more* additional cancer cases in California alone. Moreover, these risk estimates are for the "average" person who breathes less than the statewide outdoor average concentration levels of diesel exhaust. People who are exposed to higher than average levels of diesel exhaust, such as urban residents, people living near major roads, distribution centers and other diesel "hot spots," and occupationally exposed individuals, would have higher risks of lung cancer from diesel. These estimates indicate the magnitude of the task before us in reducing the diesel risk and only hint at the enormous human tragedy due to diesel exposure. Lung cancer has a poor prognosis; the five-year survival rate is less than 14 percent. Thus if 350 excess lung cancers are projected per million people exposed, 300 of these victims would likely die within five years.

I. J L Mauderly Environ Health Perspect. 1994 October; 102(Suppl 4): 165–171. Toxicological and epidemiological evidence for health risks from inhaled engine emissions.nih.gov Experimental human exposures to DE show that lung inflammatory and other cellular effects can occur after single exposures, and sparse data suggest that occupational exposures might affect respiratory function and symptoms. Epidemiology suggests that heavy occupational exposures to exhaust probably increase the risks for mortality from both lung cancer and noncancer pulmonary disease.

J.Garshick et al. (1988).To confirm the results of the case-control study, Garshick et al. (1988) evaluated the risk of lung cancer as a result of exposure to diesel exhaust from railroad locomotives. These investigators conducted a retrospective cohort study of

55,407 white male railroad workers who were aged 40 to 64 in 1959 and who had started railroad service 10 to 20 years earlier. The cohort was traced until the end of 1980. Death certificates were obtained for 88% of 19,396 deaths, and 1,694 lung cancer cases were identified. Records of yearly job assignments obtained through the Railroad Retirement Board served as an index of diesel exhaust exposure. Workers were considered to be either exposed or unexposed to diesel exhaust, depending on the yearly job code. These classifications were confirmed by measurements of current exposures to respirable particulate matter for workers in selected jobs. These measurements were analyzed as described in the earlier discussion of the 1987 case-control study (Garshick et al. 1987a). A proportional hazards model and directly standardized rates were used to calculate the relative risk of lung cancer for a worker whose job involves diesel exhaust exposure. The group of workers aged 40 to 44 in 1959 had a relative risk of 1.45 (95% CI=1.11-1.89) for lung cancer. This group consisted of workers with the longest possible duration of diesel exposure.

To control for the confounding effects of asbestos exposures in the cohort, the relative risk of lung cancer was not considered for groups of workers with possible exposures to asbestos in the past (shop workers and hostlers). When this analysis was conducted, the relative risk for lung cancer remained elevated at 1.57 (95% CI=1.19-2.05) in the group aged 40 to 44 in 1959, and it was 1.34 (95% CI=1.02-1.76) in the workers aged 45 to 49 in 1959. These results confirmed those obtained with the proportional hazards regression model.

The effects of cigarette smoking could not be eliminated because of the retrospective nature of the study. However, the prevalence of cigarette smoking was the same for workers with and without potential diesel exhaust exposure in a group of 517 current railroad workers who were surveyed in 1982 regarding past asbestos exposure (Garshick et al.1987b).

Epidemiologic studies of lung cancer risk in diesel-exposed workers are inherently problematic because of (1) the difficulty in defining and quantifying exposure, (2) the relatively short time between initial exposures and analysis of risk in some studies, and (3) the need to control for cigarette smoking. The reports by Garshick et al. (1987a;

1988) are the most thorough epidemiologic studies conducted to date. Data on cigarette smoking were collected in the case-control study. An attempt was made to control for the confounding exposure to asbestos. Attempts were also made to characterize exposures to diesel exhaust through the collection of industrial hygiene data. The period between the first diesel exposure and data analysis was adequate to allow the observation of exposure-related cancer for some age groups of the cohort. The fact that the findings of the two studies were both independent (the two studies based their analyses on different lung cancer deaths) and consistent fortifies the conclusion that occupational exposure to diesel exhaust is associated with an increased risk of lung cancer.

The studies of Garshick et al. (1987a; 1988) are subject to a number of limitations, some of them inherent, that preclude them from providing definitive evidence that diesel exhaust is an occupational carcinogen. Ascertainment of death certificates was incomplete in both studies (87% in the case-control study, and 88% in the retrospective cohort study). In both reports of final data, the authors presented data on lung cancer risk only for separate-age subcohorts within the study population. Though there is merit in the authors' rationale for splitting the groups by age, the risk analysis should have considered diesel exposure for the combined cohort also. The investigators attempted to characterize exposures to diesel exhaust by collecting industrial hygiene data, but they were forced to use an experimental approach to collect them. Exposure to diesel exhaust is difficult to measure because of the complex nature of the exhaust. Measuring exposure to respirable particulate matter as a surrogate for diesel exhaust allows for a substantial error in classification of exposures, as there is no way to define the source of the particulates. Adjusting the measurements to exclude the contribution of cigarette smoke particulates eliminates only one extraneous source of respirable particulates. The classification of exposed and unexposed workers is particularly important to the outcome of the case-control study because the unexposed workers were used as the referent population. Furthermore, no attempts were made to control for potentially confounding exposures to pyrolysis products of fuels that were used to power locomotives before the use of diesel fuel. (NIOSH.1988)

K. Garshick et al. (1987a). Garshick in 1987a conducted a case-control study of deaths among U.S. railroad workers to test the hypothesis that lung cancer is associated with exposure to diesel exhaust. The study included only male railroad workers who had at least 10 years of railroad service, were vested in the railroad retirement program, were born on or after January 1, 1900, and died between March 1, 1981, and February 28, 1982. The investigators collected death certificates for 87% of the 15,059 deaths reported to the U.S. Railroad Retirement Board. Within the cohort, 1,256 workers who died from lung cancer were matched with two deceased comparison workers by age (± 2.5 years) and date of death (± 31 days). Deceased workers whose jobs had involved exposure to diesel exhaust (engineers and firemen, brakemen and conductors, diesel locomotive repair workers, and hostlers) were compared with deceased workers without occupational exposure to diesel exhaust (clerks and station agents).

Work histories were determined from yearly job reports filed with the U.S. Railroad Retirement Board. These reports were used to classify workers as exposed or unexposed to diesel exhaust. The classifications were confirmed by measuring current exposures to respirable particulate matter for workers in selected jobs. Respirable particulate matter was chosen as a marker for exposure to the particulate fraction of diesel exhaust. The respirable particulate fraction was sampled because it included all of the diesel exhaust particulates and excluded some of the larger nondiesel particulates. Respirable dust exposures were corrected for cigarette smoke particulates by analyzing the nicotine content of composite samples (Woskie et al. 1988a; Woskie et al. 1988b). An adjusted respirable particulate concentration was then calculated for each job group by subtracting the applicable average fraction of cigarette smoke from each railroad's average respirable particulate concentration. Personal exposure to respirable particulate matter was measured for 39 common jobs in four U.S. railroads over a 3-year period.

Table a . Characteristics of Epidemiologic Studies of Exposure to Diesel Exhaust and Carcinogenicity, Published Since 1986

Investigator	Population Studied	Observation Period	Findings	Comments
Garshick et al. 1987a	U.S. railroad workers born in 1900 or later with 10 or more years of service	1959-81 for diesel exhaust exposure; deaths that occured between March 1, 1981, and February 28, 1982.	Workers exposed occupationally to diesel exhaust for 20 years had a significantly increased relative odds ratio (1.41, 95% CI*=1.06, 1.88) of lung cancer	Population-based, case-control study that included industrial hygiene characterization of exposures and multiple conditional logistic regression analysis to adjust for confounders such as smoking and asbestos exposures. Only 87% of death certificates were collected.
Garshick et al. 1988	U.S. railroad workers aged 40 to 64 in 1959 who started railroad service 10 to 20 years earlier	1959-1980 for diesel exposure; deaths that occurred before December 31, 1980	Workers aged 40-44 in 1959 had a significantly increased relative risk (1.45, 95% CI=1.11, 1.89) of lung cancer	Retrospective cohort study. Only 88% of death certificates were collected. Effects of smoking could not be eliminated. The effect of asbestos exposures was addressed by considering diesel-exposed workers separately from asbestos-exposed workers using a proportional hazards

				regression model.

*CI= confidence interval. [return to table]

According to the authors, diesel locomotives replaced steam locomotives over a short period (from 14% diesel use in 1947 to 95% in 1959). Thus the year 1959 was chosen as the effective beginning of diesel exhaust exposure for this study. Workers who retired before that year were classified as unexposed to diesel exhaust. The authors acknowledged that some workers had additional earlier years of diesel exhaust exposure. Smoking histories were obtained by questionnaires from the deceased workers' closest relatives or by direct telephone contact if there was no response to the questionnaire. Asbestos exposures in railroad workers occurred primarily in the steam engine era. Asbestos exposure for this study was therefore categorized by the job held in 1959 (the end of the steam locomotive era) or by the last job held if retirement occurred before 1959.

The relative hazard of lung cancer attributable to diesel exhaust exposure was calculated using a multiple conditional logistic regression to adjust for smoking and asbestos exposure. A statistically significant increase in relative odds (1.41, 95% CI=1.06-1.88) was found for lung cancer among workers aged 64 or younger at the time of death who had worked in a [diesel-exposed] job with diesel exposure for 20 years. No increase was found in workers aged 65 or older. The authors felt that this finding reflected the fact that many of these men retired shortly after the transition to diesel-powered locomotives. (NIOSH.1988)

L. Mauderly et al. (1987).Mauderly et al. (1987) reported the results of a carcinogenicity study in which F344 male and female rats were exposed to unfiltered diesel exhaust at three concentrations for up to 30 months. Diesel exhaust was generated by stationary light-duty diesel engines (1980 model, 5.7-liter, Oldsmobile V-8) operated by computer through continuously repeating U.S. Federal Test Procedure urban certification cycles. The exhaust effluents were diluted 10:1 with filtered air, serially diluted to the final concentrations, and then directed through exposure chambers. The average particulate concentrations for the low medium, and high exposures to diesel exhaust were 0.35, 3.5, and 7 mg/m³, respectively. Concentrations of key components identified in the diesel exhaust are shown in Table b.

Male and female F344 rats that were 15 weeks old were randomly divided by litter into four treatment groups. There were 365 rats in the control group, and 366, 367, and 364 rats in the low-, intermediate-, and high-exposure groups, respectively. Rats were added to all treatment groups during February 1981; a second group of rats was added to all treatment groups 1 year later. Both groups of rats were derived from the same breeding colony and were exposed in the same chambers. The two added groups of rats were treated as one study population since the groups showed no differences in body weights or survival times. All animals were exposed to unfiltered diesel exhaust for 7 hours per day, 5 days per week for up to 30 months.

All rats terminated for histopathology and all rats that died or were euthanized received a complete necropsy. All lesions except for soot macules and representative portions of each lung lobe were examined microscopically, as were samples of other respiratory tract tissues. Exposure to diesel exhaust did not cause overt signs of toxicity. No significant differences in body weight or life span were observed in either the males or females in the experimental groups compared with the controls. The physical condition of all groups of animals appeared to be similar.

Table b. Concentrations of Key Components of Exposure Atmospheres* (Adapted from Mauderly et al., 1987)

Component	Control	Low	Medium	High
Particulate (mg/m³)	0.010 (0.010)†	0.350 (0.070)	3.470 (0.450)	7.80 (0.810)
Carbon monoxide (ppm)	1 (1)	3 (1)	17 (7)	30 (13)
Nitric Oxide (ppm)	0	0.6 (0.3)	5.4 (1.5)	10.0 (2.6)
Nitrogen dioxide (ppm)	0	0.1 (0.1)	0.3 (0.2)	0.7 (0.5)
Hydrocarbons (ppm)	3 (1)	4 (1)	9 (5)	13 (8)

| Carbon dioxide (%) | 0.2 (0.04) | 0.2 (0.03) | 0.4 (0.06) | 0.7 (0.1) |

*Mean of weekly mean values during 30 months of exposure. [return to table]
†Figures in parentheses are standard error.

Soot accumulated progressively and significantly in the lungs of all exposed rats. After 24 months of exposure, the mean lung burden of diesel exhaust particulate per rat was reported to be 0.6 ± 0.02 mg for the low exposure, 11.5 ± 0.5 mg for the intermediate exposure, and 20.8 ± 0.8 for the high exposure. These calculations were made by measuring the amount of light absorbed by lung homogenates from exposed rats and comparing those values with standard curves constructed from measurements made from known amounts of soot deposited in the lungs of unexposed rats.

Changes in the epithelial lining of the air spaces and progressive fibrosis occurred in the areas of soot accumulation. Hyperplasia and squamous metaplasia were seen in broncho-alveolar spaces.

Broncho-alveolar adenomas, adenocarcinomas, benign squamous cysts, and squamous cell carcinomas were observed in the lungs of exposed rats. Rats exposed to the high concentration of diesel exhaust for up to 30 months experienced statistically significant increases in adenocarcinomas, benign squamous cell tumors, and squamous cell carcinomas in male and female rats. The increase in total tumor incidence for the high-exposure group was also statistically significant when compared with the control group. The percentages of rats with lung tumors (males and females combined) are listed by exposure group in Table c.

Table c. Percentages of Rats (Male and Female Combined) with Lung Tumors, by Exposure Group (Adapted from Mauderly et al., 1987)

Exposure group	Adenomas	Adenocarcinomas and squamous cell carcinomas	Squamous cysts only	All tumors
High	0.4	7.5*	4.9*	12.8*

Medium	2.3*	0.5	0.9	3.6*
Low	0	1.3	0	1.3
Control	0	0.9	0	0.9

* Significantly higher than controls at p<0.05 by z statistic.

A statistically significant increase in adenomas occurred in the intermediate-exposure group. One adenocarcinoma and two squamous cysts were also observed in female animals in that group. The increase in all tumors in the intermediate-exposure group was statistically significant when compared with the control group.

No statistically significant increases in tumor incidence occurred in the low-exposure animals. Squamous tumors were always associated with focal areas of engine soot retention, epithelial cell alterations, and fibrosis. They are thus likely to represent a progression of squamous metaplasia. These tumors may have resulted from a generalized response to the accumulation of relatively insoluble particles.

None of the lung tumors had metastasized to pulmonary lymph nodes or other organs. Increased numbers of adducts were found in DNA extracted from the lungs of rats exposed to the highest concentration of diesel exhaust for 30 months.

Diesel exhaust inhaled chronically at the intermediate and high concentrations in this study induced a significant number of benign and malignant pulmonary tumors in male and female rats. The increased numbers of DNA adducts suggest that tumor development may have been initiated by the interaction of reactive metabolites of soot-associated organic compounds with lung cell DNA. In this study, the relationship between lung burden of diesel exhaust particulates and tumor prevalence was progressive rather than linear with time, rising rapidly late in the exposure regimen. (NIOSH.1988)

L.Heinrich et al. (1986).Results of an extensive long-term inhalation study of cancer in mice, rats, and hamsters exposed to filtered and unfiltered light-duty diesel engine exhaust were reported by (Heinrich et al. 1986). Equal numbers of male and female

Syrian golden hamsters, female NMRI mice, and female Wistar rats were exposed to clean air, unfiltered diesel exhaust, or filtered diesel exhaust. Each group consisted of 96 animals. All experimental animals were 8 to 10 weeks old at the start of the study. Throughout the lifespan of the animals, exposure was for 19 hours per day, 5 days per week. The maximum duration of exposure was 120 weeks for hamsters and mice, and 140 weeks for rats.

A stationary 1.6-liter diesel engine operated according to EPA's US-72 driving cycle was used to generate the exhaust. The diesel fuel used was a European reference fuel with a sulfur content of 0.36%. The exhaust was diluted with filtered air to a volume rate of 1:17 (diesel exhaust/air) and was then directed into an exposure chamber. At this dilution rate, the measured concentration of the particulate fraction of diesel exhaust was approximately 4 mg/m^3. To remove particulates from the exhaust, the diesel emissions were passed through a centrifugal separator and/or a particle filter. The concentrations of exhaust components in the inhalation chambers for both filtered and unfiltered exhaust are shown in Table e.

Table d. Characteristics of recent studies* of carcinogenicity in animals exposed to diesel exhaust by inhalation

Study	Type of engine	Nature of exhaust	Animal species	Exposure time	Particulate exposure concentrations (mg/m^3)	Findings
Heinrich et al. [1986]	1.6-liter Volkswagen	Unfiltered	Female Wistar rats	19 hr/day, 5 days/week, max. of 140 weeks	4	Significantly increased incidence of adenomas, benign squamous cell cysts, and squamous cell carcinoma of the

						lung when compared with controls.
		Filtered	Female Wistar rats	19 hr/day, 5 days/week, max. of 140 weeks	----	No significant differences in histopathological findings compared with controls.
		Unfiltered	Female NMRI mice	19 hr/day, 5 days/week, max. of 120 weeks	4	Statistically significant increase in malignant and total lung tumors (because of increases in adenocarcinomas) when compared with controls.
		Filtered	Female NMRI mice	19 hr/day, 5 days/week, max. of 120	----	Statistically significant increase in malignant and total lung tumors (because of increases in adenocarcinomas) when compared with controls.

		Unfiltered	Male and female Syrian golden hamsters	19 hr/day, 5 days/week, max. of 120 weeks	4	No significant differences in histopathological findings compared with controls
		Filtered	Male and female Syrian golden hamsters	19 hr/day, 5 days/week, max. of 120 weeks	----	No significant differences in histopathological findings compared with controls.
Mauderly et al. [1987]	5.7-liter Oldsmobile	Unfiltered	Male and female F344 rats	7 hr/day, 5 days/week, max. 30 months	0.35 3.5 7.0	High exposure led to statistically significant increases in benign squamous cysts and malignant tumors (adenocarcinomas and squamous carcinomas) compared with controls. Intermediate exposure led to a statistically

						significant increase in adenomas and total tumors when compared with controls. There were no statistically significant increases in benign or malignant tumors in low-exposure animals.
Brightwell et al. [1986]	1.5-liter Volkswagen	Unfiltered	Male and female F344 rats	16 hr/day, 5 days/week, 2 years†	0.7 2.2 6.6	Statistically significant increase in undefined tumors in both male and female high-exposure animals compared with controls. Statistically significant increase in undefined tumors in female intermediate-

						exposure animals compared with controls.
		Filtered	Male and female F344 rats	16 hr/day, 5 days/week, 2 years†	Below the limit of detection	No increase in tumor incidence in any exposure group when compared with controls.
		Unfiltered	Male and female syrian hamsters	16 hr/day, 5 days/week, 2 years	0.7 2.2 6.6	No increase in tumor incidence in any exposure group when compared wth controls.
		Filtered	Male and female Syrian hamsters	16 hr/day, 5 days/week, 2 years	Below the limit of detection	No increase in tumor incidence in any exposure group when compared with controls.
Ishinishi et al. [1986]	Light-duty, 1.8-liter, 4-cylinder, swirl chamber	Unfiltered	Male and female F344 rats	16 hr/day, 6 days/week, max. 30 months	0.1 0.4 1 2	No statistically significant increase in lung tumors. Increase in hyperplasia, squamous hyperplasia,

						interstitial fibrosis, hyperplastic lesions.
	Heavy-duty, 11-liter, 6-cylinder, direct injection	Unfiltered	Male and female F344 rats	16 hr/day, 6 days/week, max. 30 months	0.4 1 2 4	Statistically significant increase in lung tumors (adenoma, squamous cell carcinoma, adenocarcinoma, adenosquamous carcinoma) in the high-exposure group compared to controls.
		Filtered	Male and female F344 rats	16 hr/day, 6 days/week, max. 30 months	0.005 0.019	No statistically significant increase in lung tumors. Increase in hyperplasia, squamous hyperplasia, interstitial fibrosis, hyperplastic lesions.
Iwai et al., 1986	2.4-liter	Unfiltered	Female F344 rats	8 hr/day, 7 days/week, 24	4.9	Statistically significant increase in total

				months§		lung tumors (adenomas, adenocarcinoma, adenosquamous carcinomas, squamous carcinoma, and large cell carcinoma) compared with controls. Statistically significant increase in splenic malignant lymphoma compared with controls. Increased incidence of tumors other than lung, and multiplicity of tumors.
		Filtered	Female F344 rats	8 hr/day, 7 days/week, 24 months§	----	Minimal histopathologic changes. Statistically significant increase in

						splenic malignant lymphoma compared with controls. Increased incidence of tumors other than lung, and multiplicity of tumors

*Since 1986

† period was 30 months for surviving animals

§ period was 30 months for some animals.

In addition, some diesel-particle-associated PAHs were measured in the unfiltered exhaust. Concentrations of 13 ng/m^3 of benzo(a)pyrene, 21 ng/m^3 of benzo(e)pyrene, and 51 ng/m^3 of a mixture of benzofluoranthenes were measured in batched samples. The authors did not specify the number of samples or the analytical method used to determine these concentrations. Interpretation of the hamster data is complicated by the fact that the animals were treated with antibiotics for several months during the study. Control hamsters were also treated with antibiotics. No significant differences in body weight developed between control and exposed hamsters over the entire length of the study. In contrast, mice and rats exposed to unfiltered diesel exhaust showed decrements in body weight after approximately 480 days. The median survival time of animals was not affected by exposure. Tissues from the nasal cavity, sinuses, larynx, trachea, esophagus, lungs, forestomach, glandular stomach, liver, kidneys, adrenals, and urinary bladder were subjected to histopathologic examination. In some cases, the salivary glands, thyroid, thymus, aorta, heart, spleen, lymph nodes, and ovaries were also subjected to histopathologic examination.

Histopathology of hamsters exposed to diesel exhaust failed to demonstrate the induction of tumors in the lung or upper respiratory tract. Significant deposits of soot particles were evident in the hamsters exposed to unfiltered diesel exhaust. The lungs of these animals exhibited an increased incidence (measured qualitatively) of thickened septa, bronchioalveolar hyperplasia, and emphysema compared with controls. No significant differences were found between the controls and the animals exposed to filtered diesel exhaust.

Mice exposed to filtered or unfiltered diesel exhaust showed a 2.5-fold increase in lung tumor incidence compared with controls. Combined benign and malignant tumor incidences were as follows: 13% in the control group at the end of the study; 31% in the group exposed to filtered diesel exhaust; and 32% in the group exposed to unfiltered diesel exhaust. This increase was predominantly due to the induction of adenocarcinomas. Bronchioalveolar hyperplasia was more frequent (64%) in the group of mice exposed to unfiltered diesel exhaust than in mice exposed to filtered diesel exhaust (15%) and controls (5%). Furthermore, multifocal alveolar lipoproteinosis was found in 71% of the mice exposed to unfiltered diesel exhaust compared with 3% for filtered diesel exhaust and 4% for controls. Similar results were obtained for interstitial fibrosis that occurred almost exclusively in the mice exposed to unfiltered diesel exhaust.

Of the 95 rats exposed to unfiltered diesel exhaust, 15 exhibited a total of 17 lung tumors. These tumors were classified as 8 bronchiolo-alveolar adenomas and 9 squamous cell tumors (8 benign keratinizing cysts and 1 squamous cell carcinoma). Hyperplasia was seen in the lungs of 94 of the 95 rats exposed to unfiltered diesel exhaust, and metaplasia occurred in 62 of these animals. Other severe inflammatory changes such as thickened septa, foci of macrophages, and cholesterol crystals were found in the lungs of rats exposed to unfiltered diesel exhaust. No changes were seen in the control animals or in those exposed to filtered diesel exhaust. No exposure-related changes were observed in the upper respiratory tracts of the rats exposed to unfiltered diesel exhaust.

Exposure to filtered and unfiltered diesel exhaust resulted in a statistically significant increase in the incidence of lung adenocarcinomas in female NMRI mice. In hamsters, long-term exposure to unfiltered diesel exhaust led to broncho-alveolar hyperplasia and emphysematous lesions in the respiratory tract, but it did not produce tumors. In rats, long-term exposure to unfiltered diesel exhaust led to extensive hyperplasia and metaplasia of the broncho-alveolar epithelium and to a significantly increased incidence of adenomas and squamous cell tumors of the lung compared with controls. (NIOSH.1988)

Table e. Concentrations of Exhaust Components for Filtered and Unfiltered Diesel Exhaust Measured in the Exposure Chambers (mean ± standard deviation) (Adapted from Heinrich et al., 1986)

Component	Control (clean air)	Filtered exhaust	Unfiltered exhaust
Carbon monoxide (ppm)	0.16 ± 0.27	11.1 ± 1.92	12.5 ± 2.18
Carbon dioxide (vol.%)	0.10 ± 0.01	0.35 ± 0.05	0.38 ± 0.05
Sulfur dioxide (ppm)	----	1.02 ± 0.62	1.12 ± 0.89
Oxides of nitrogen (ppm)	----	9.9 ± 1.80	11.4 ± 2.09
Nitric oxide (ppm)	----	8.7 ± 1.84	10.0 ± 2.09
Nitrogen dioxide (ppm)	----	1.2 ± 0.26	1.5 ± 0.33
Alkanes (ppm)	3.5 ± 0.29	5.2 ± 0.65	5.5 ± 0.69
Methane (ppm)	2.3 ± 0.17	2.4 ± 0.20	2.6 ± 0.19
Alkanes without methane (ppm)	1.3 ± 0.5	2.9 ± 0.50	3.1 ± 0.53
Particles (mg/m^3)	----	----	4.24 ± 1.42

M.Brightwell et al. (1986).Brightwell et al. (1986) reported preliminary data on the chronic toxicity of diesel engine exhaust in Fischer 344 rats and Syrian hamsters. Each group of experimental animals was made up of 144 rats and 312 hamsters with approximately equal numbers of males and females. The animals were 6 to 8 weeks old at the start of the 2-year exposure period.

The diesel exhaust emissions used in this study were generated by a 1.5-liter light-duty diesel engine. The emissions were diluted to yield particulate exposure concentrations of approximately 0.7, 2.2, and 6.6 mg/m^3. The diesel emissions were either subjected to particle filtration or they were unaltered (unfiltered). Exposures were carried out overnight for 16 hours per day, 5 days per week.

The mean concentrations for carbon monoxide (CO) and nitrogen oxides (NO_x) for both filtered and unfiltered diesel exhaust in the high-exposure groups were 32 and 8 parts per million (ppm), respectively. Concentrations of those contaminants for the controls were 1 ppm CO and 0.1 ppm NO_x. No data were presented for the concentrations of these contaminants in the intermediate- and low-exposure chambers. Similarly, no data were presented for exposure concentrations of other exhaust components, although regular analyses were conducted for particle size distributions, aldehydes, phenols, PAHs, sulfates, and individual hydrocarbons.

Interim sacrifices of rats were conducted after 6, 12, 18, and 24 months of exposure, while hamsters had interim sacrifices at 6 and 16 months. All surviving hamsters were sacrificed at the end of 2 years of exposure. Rats that survived 2 years of exposure were maintained for up to 6 additional months without further exposure to exhaust.

Animals that died or were sacrificed were subjected to a full necropsy, and histopathology was carried out on the respiratory tracts (nasal passages, larynx, trachea, and lungs) of all animals in the high-exposure and control groups. Histologic examinations were also performed on the respiratory tracts of all animals in the groups with intermediate and low exposures to filtered and unfiltered diesel exhaust. Histopathology was carried out on all suspected tumors regardless of experimental treatment.

The major histopathologic finding in the study was an increase in the incidence of primary lung tumors in rats exposed to the intermediate and high concentrations of unfiltered diesel exhaust. Table 7 summarizes the histopathologic findings for primary benign and malignant lung tumors in rats exposed to unfiltered diesel exhaust and to air alone. The incidence of primary lung tumors was 2%, 1%, 4%, and 23% for male rats in the control, low-, intermediate-, and high-exposure groups, respectively. Lung tumor incidence in female rats was 1%, 0%, 15%, and 54% for control and for low-, intermediate-, and high-exposure groups, respectively. However, a later analysis of these data (Fouillet and Brightwell 1987) points out that tumor incidence was based on the total number of animals examined in each treatment group. Since this total included some animals sacrificed after 6, 12, 18, and 24 months of exposure, some of them clearly were not exposed long enough to induce recognizable lung tumors. When tumor incidence was recalculated using only the data for rats surviving beyond 24 months, 44% of male and 99% of female rats exposed to unfiltered diesel exhaust developed lung tumors.

Table f. Primary Benign and Malignant Lung Tumors in Rats Exposed to Unfiltered Diesel Exhaust* (Adapted from Brightwell et al., 1986)

		F344 rats that developed lung tumors			
Exposure group	Particulate Concentration (mg/m^3)	Males Number	%	Females Number	%
High	6.6	16/71†	23	39/72†	54
Intermediate	2.2	3.72	4	11/72†	15
Low	0.03	1/72	1	0.72	0
Control	0	3/140	2	1/142	1

*These figures represent calculations that included animals sacrificed after 6, 12, 18, and 24 months of exposure.

† significant compared with controls ($p<0.01$).

No increase occurred in the incidence of primary lung tumors in any other treatment group. Respiratory tract tumors were rare in hamsters and were not attributed to treatment.

The pathology description in this report [Brightwell et al. 1986] is very limited. It contains no specific diagnoses of the lung tumors, and no data on whether the tumors were single, multiple, lethal, or incidental. Data on degree of invasion are also absent. No comment is made on pathology of the nasal passages, larynx, or trachea. There is no description or discussion of chronic toxicity, hyperplasia, or the relationship of hyperplasia to neoplasia. Although the investigators did not present the full spectrum of their bioassay data, the information presented justifies the conclusion that long-term exposure to high concentrations of unfiltered diesel exhaust leads to a significant increase in the incidence of benign and malignant lung tumors in male and female F344 rats .(NIOSH.1988)

N. Ishinishi et al. (1986). Ishinishi and coworkers in 1986 studied F344 rats to determine the effects of long-term inhalation of exhausts from heavy- and light-duty diesel engines. Five-week-old female and male F344 rats were exposed to various concentrations of diesel exhausts for 6, 129 18, 24, and 30 months, 16 hours per day, 6 days per week. Three types of exposures were administered: exposure to filtered and unfiltered diesel exhaust from 11-liter heavy-duty engines, and exposure to unfiltered diesel exhaust from 1.8-liter light-duty engines.

For the carcinogenicity experiment, five groups of animals (each consisting of 64 male and 59 female rats) were exposed for 30 months to unfiltered heavy-duty engine exhaust at a given particulate concentration originally designed to be 0, 0.4, 1, 2, and 4 mg/m^3 (see Table g for actual concentrations). Five additional groups (each consisting of 64 male and 54 female rats) were exposed for 30 months to unfiltered light-duty engine exhaust at a given particulate concentration originally designed to be 0, 0.1, 0.4, 1, and 2 mg/m^3 (see Table g for actual concentrations).

Two additional groups of 64 male rats were exposed to filtered exhaust from the heavy-duty engines. The 0.4- and 4-mg/m^3 concentrations were filtered so that the animals were exposed only to the gaseous fraction of the exhaust. For comparison, three additional groups of 64 male rats were exposed to the unfiltered diesel exhaust from the same source at particulate concentrations of 0, 0.4, and 4 mg/m^3. Table (g) presents a summary of gas and particle concentrations for each exposure atmosphere.

A concentration-dependent decrease in body weight was observed, with the greatest effect observed in the 4-mg/m^3 group exposed to exhaust from heavy-duty engines.

Table g. Summary of Gas and Particle Concentrations*(Adapted from Ishinishi et al., 1986)

Type of diesel exhaust (%)	Particle concentration(mg/m)^3		Gaseous Concentration								
			NO_x (ppm)	NO (ppm)	NO_2 (ppm)	CO (ppm)	CO_2 (%)	SO_2 (ppm)	Formal-dehyde (ppm)	SO_4^{-2}	O_2 (%)
	Target†	Actual									
Unfiltered exhaust from heavy duty engine	4	3.72	37.45	34.45	3.00	12.91	0.360	4.57	0.29	361	20.4
	2	1.84	21.67	19.99	1.68	7.75	0.215	2.82	0.18	198	20.6
	1	0.96	13.13	12.11	1.02	4.85	0.140	1.79	0.11	111	20.7
	0.4	0.46	6.17	5.71	0.46	2.65	0.084	0.98	0.05	62.9	20.8

	0	0.002	0.061	0.042	0.021	0.63	0.035	0.06	0.003	0.49	20.8
Filtered and unfiltered exhaust from heavy-duty engine	4 (unfiltered)	2.99	36.45	31.50	4.95	12.90	0.412	4.03	0.20	358	20.3
	4 (filtered)	0.019	36.76	32.81	3.96	13.00	0.391	4.50	0.24	1.61	20.4
	0.4 (unfiltered)	0.39	5381	5.37	0.44	2.50	0.084	0.98	0.04	57.7	20.7
	0.4 (filtered)	0.005	5.58	5.16	0.472	2.54	0.083	0.96	0.04	1.43	20.7
Unfiltered exhaust from light-duty engine	0	0.004	0.062	0.040	0.024	0.06	0.068	0.03	0.003	0.35	20.8
	2	2.32	20.34	18.93	1.41	7.10	0.418	4.70	0.13	315	20.3
	1	1.08	10.14	9.44	0.70	3.96	0.219	2.42	0.07	151	20.5
	0.4	0.41	4.06	3.81	0.26	2.12	0.105	1.06	0.03	62.4	20.7
	0.1	0.11	1.24	1.16	0.08	1.23	0.050	0.38	0.01	18.8	20.8

*Exposures were for 30 months.   [return to table]

† are identified in the text by the design exposure.

After 6 months of exposure, "anthracosis"* was observed in all groups exposed to particle-containing exhausts. Severity was proportional to concentration and duration of exposure. Hyperplasia of type II epithelial cells and bronchiolar epithelium associated with anthracosis was observed after 18 months in the groups exposed to the higher particle concentrations. The extent of these conditions depended on exposure. Squamous metaplasia with focal interstitial fibrosis was often observed in hyperplastic lesions of the subpleural zone. Scanning electron microscopy revealed irregularity, shortening, and absence of cilia in the mucosal epithelia of the trachea and main bronchi. These lesions were also observed in rats exposed only to the gaseous components of the exhaust; the severity of the lesions increased in proportion to exhaust concentration and duration of exposure.

A statistically significant increase occurred in the incidence of lung tumors in rats (male and female combined) exposed for 30 months to heavy-duty diesel engine exhaust with particulate concentrations of 4 mg/m^3 compared with controls. Tumor incidence was 6.5% (8/124) for exposed rats and 0.8% (1/123) for controls. The majority of tumors were squamous cell carcinomas, adenosquamous carcinomas, and adenocarcinomas. (NIOSH.1988)

O. Iwal et al, (1986). Iwai et al.(1986) studied the effects of long-term inhalation exposure to filtered and unfiltered diesel exhaust in female F344 rats. Exhaust for this experiment was generated by a 2.4-liter light-duty diesel engine. Seven-week-old female rats (initially 24 animals/exposure group) were exposed to unfiltered exhaust, filtered exhaust, or clean air for 24 months, 8 hours/day, 7 days/week. Measured concentrations (mean ± standard deviation) of exhaust components were 4.9 ± 1.6 mg/m^3 for-particles, 30.9 ± 10.9 ppm for oxides of nitrogen (NO_x), 1.8 ± 1.8 ppm for nitrogen dioxide (NO_2), 13.1 ± 3.6 ppm for sulfur dioxide (SO_2), and 7.0 ± 1.4 ppm for carbon monoxide (CO). Some animals were autopsied after 3, 6, 12, and 24 months of exposure. Histologic and electron-microscopic examinations were performed on lungs, spleen, and other organs. Some of the rats exposed for 12 to 24 months were kept in clean air for an additional 3 or 6 months and then examined.

After 6 months of exposure to unfiltered diesel exhaust, phagocytotic macrophages filled with black particles were distributed in an irregular pattern in the lungs. Areas where macrophages were gathered showed proliferations of Type II alveolar epithelial cells showing adenomatous metaplasia. More lesions were found after 1 year of exposure, but no neoplastic lesions were observed. Two adenomas were found in one of five rats kept in clean air for 3 months after 1 year of exposure to unfiltered exhaust. After 2 years of exposure, the number of particles in the macrophages increased markedly. Fibrous thickening of alveolar walls was observed, and mast cell infiltration was found with epithelial hyperplasia where macrophages gathered. Neoplastic changes were observed after 2 years of exposure; some of these showed intra-lymphatic invasion indicative of malignant transformation. Two types of lung carcinoma (adenocarcinoma and squamous or adenosquamous carcinoma) were observed.

In the rats exposed to filtered diesel exhaust, histologic changes were minimal, and no heterotrophic hyperplasia was observed in the alveolar walls. Quantitative analysis of epithelial proliferative changes in the lung indicated an increase in affected areas that was associated with the length of exposure to unfiltered exhaust.

Rats exposed to unfiltered exhaust had a statistically significant increase in lung tumor incidence compared with controls. After 24 months of exposure, 4/14 rats had tumors, 2 of which were malignant. After an additional 6 months in clean air, four of the five remaining rats had tumors, three of which were malignant. The combined incidence of tumors was 42% (8/19) in rats exposed to unfiltered diesel exhaust compared with 0% (0/16) in rats exposed to filtered exhausts and 4.5% (1/22) in the controls. The distribution of tumor types found in rats exposed to unfiltered diesel exhaust was as follows:

Tumor Type	# of Rats Exposed
Adenomas	3 rats
Adenocarcinoma	3 rats
Adenosquamous carcinomas	2 rats
Squamous carcinoma	1 rat
Large cell carcinoma	1 rat

The tumor found in the control rat was an adenoma. The authors concluded that the significantly higher incidence of lung tumors in the unfiltered exhaust exposure group could be attributed to the inhalation of particles.

Another important observation in this study was a statistically significant increase in the incidence of splenic malignant lymphoma, with or without leukemia. After 24 months, the incidence rate was 25.0% (6/24) for rats exposed to unfiltered diesel exhaust, 37.3% (9/24) for rats exposed to filtered diesel exhaust, and 8.2% (2/24) for controls. The incidence of tumors in other organs also increased in rats exposed to filtered exhaust (25%, or 6/24) and unfiltered exhaust (29%, or 7/24) compared with controls (8.2%, or 2/24). The multiplicity of tumors increased both in rats exposed to unfiltered exhaust and in those exposed to filtered exhaust, with a quadrupled incidence of tumors noted only in the unfiltered exhaust group. (NIOSH.1988)

Indoor Pollution :We are not secure enough behind the doors !

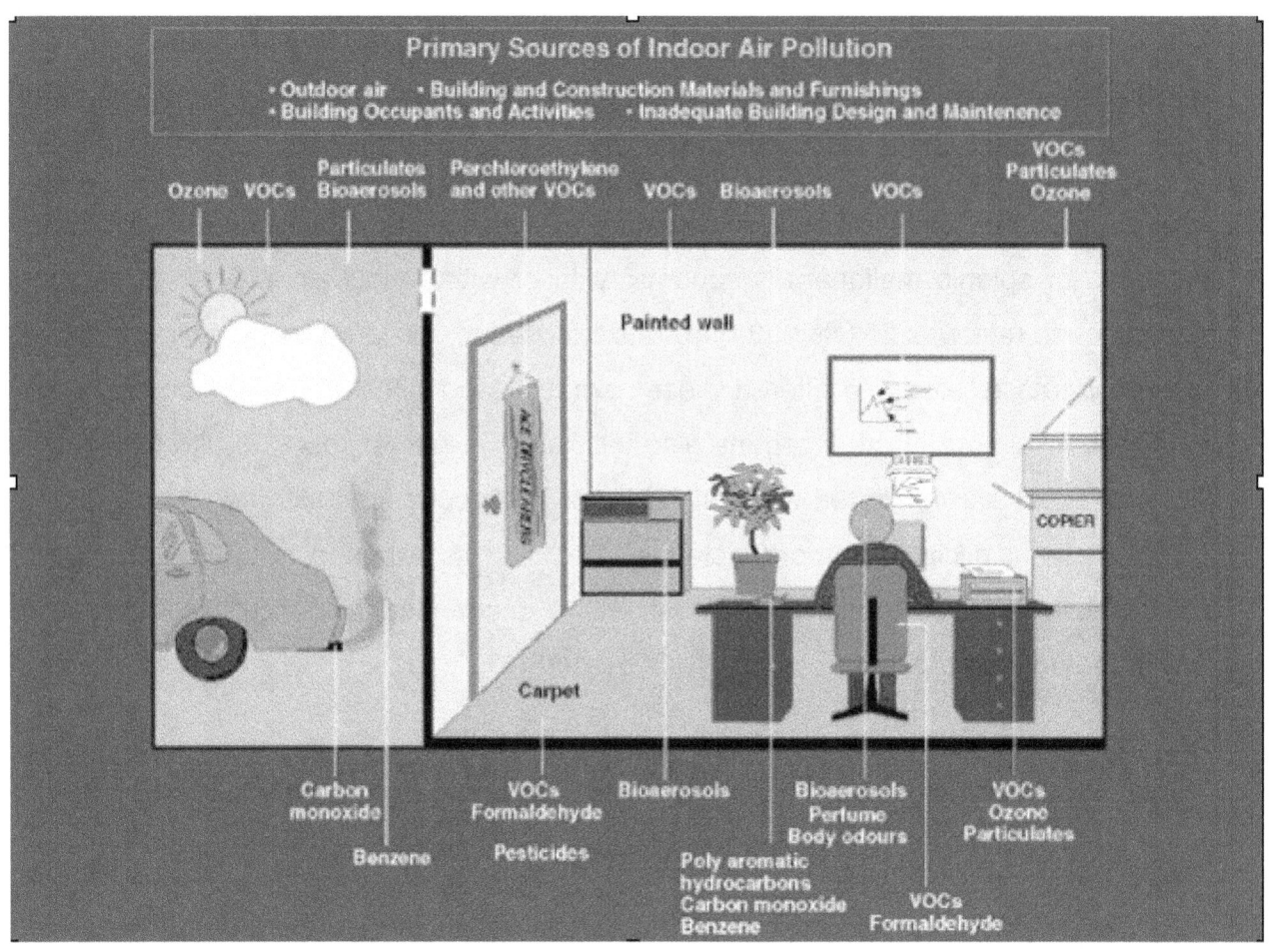

Fig I: Indoor air pollutants. *(ecourse.vtu.ac.in)*

Pollution is just does not limited to outdoor, homes are not even safe as outdoor exposures account for only a portion of risk for many compounds. It refers to the physical, chemical, and biological characteristics of air in the indoor environment within a home, building, or an institution or commercial facility (http://edugreen.teri.res.in/explore/air/indoor.htm).Indoor pollutants include particulates,aldehydes,PAH,VOC,CO etc.In two older studies using indoor concentrations from homes and offices, one by Tancrede et al. (1987) and another by McCann et al. (1986), calculated cancer potency factors with data from animal and human studies. Tancrede et al. found annual mean risks from indoor air to be about 1 in 10,000 to 1 in 100,000, and McCann et al.'s risks are about an order of magnitude higher (McCann et al. 1986; Tancrede et al. 1987).And not to forget these studies were conducted in America where environmental conditions are far better than India.

Personal exposure measurements from theTotal Exposure Assessment Methodology (TEAM) studies provided estimates of individual cancer risks from benzene ranging from 1 in 10,000 for nonsmokers to 7 in 10,000 for smokers (Wallace 1991a). Payne-Sturges et al. (2004) found risks from personal exposure over three times higher than those calculated using the ASPEN modeled outdoor concentrations. Sax et al. (2006) also found risks from personal exposures of inner-city teenagers to be on the order of 1 in 10,000.. In developing countries, the problem of indoor air pollution far outweighs the ambient air pollution.

Pollutant Sources

There are four principal sources of pollutants of indoor air 1 : (i) combustion, (ii) building material, (iii) the ground under the building, and (iv) bioaerosols.However,There are many sources of indoor air pollution in any home. These include combustion sources such as oil, gas, kerosene, coal, wood, and tobacco products; building materials and furnishings as diverse as deteriorated, asbestos-containing insulation, wet or damp carpet, and cabinetry or furniture made of certain pressed wood products; products for household cleaning and maintenance, personal care, or hobbies; central heating and cooling systems and humidification devices; and outdoor sources such as radon, pesticides, and outdoor air pollution.

The relative importance of any single source depends on how much of a given pollutant it emits and how hazardous those emissions are. In some cases, factors such as how old the source is and whether it is properly maintained are significant. For example, an improperly adjusted gas stove can emit significantly more carbon monoxide than one that is properly adjusted.

Some sources, such as building materials, furnishings, and household products like air fresheners, release pollutants more or less continuously. Other sources, related to activities carried out in the home, release pollutants intermittently. These include smoking, the use of unvented or malfunctioning stoves, furnaces, or space heaters, the use of solvents in cleaning and hobby activities, the use of paint strippers in redecorating activities, and the use of cleaning products and pesticides in housekeeping. High pollutant concentrations can remain in the air for long periods after some of these activities(http://www.epa.gov/iaq/ia-intro.html).

In developed countries the most important indoor air pollutants are radon, asbestos, volatile organic compounds, pesticides, heavy metals, animal dander, mites, moulds and environmental tobacco smoke. However, in developing countries the most important indoor air pollutants are the combustion products of unprocessed solid biomass fuels used by the poor urban and rural folk for cooking and heating. When a building is not properly ventilated, pollutants can accumulate and reach concentrations greater than those typically found outside. This problem has received media attention as "Sick Building Syndrome". Environmental tobacco smoke (ETS) is one of the main contributors to indoor pollution, as are CO, NO, and SO_2, which can be emitted from furnaces and stoves. Cleaning or remodeling a house is an activity that can contribute to elevated concentrations of harmful chemicals such as VOCs emitted from household cleaners, paint, and varnishes. Also, when bacteria die, they release endotoxins into the air, which can cause adverse health effects31. A geogenic source of indoor air pollution is radon32.(http://www.epa.gov/iaq/pubs/hpguide.html)

DISEASES ASSOCIATED WITH INDOOR AIR POLLUTANT EXPOSURE

Respiratory illness, cancer, tuberculosis, perinatal outcomes including low birth weight, and eye diseases are the morbidities associated with indoor air pollution(Bruce, N.et al.2000).

Respiratory Illness

The effect of air pollutants in general would depend on the composition of the air that is inhaled which will depend on the type of fuel used and the conditions of combustion, ventilation and duration for which the inhalation occur. The most commonly reported and obvious health effect of indoor air pollutants is the increase in the incidence of respiratory morbidity. Studies by the NIOH(NIOH.1982) on the prevalence of respiratory symptoms in women using traditional fuels (biomass) (n=175) and LPG (n=99), matched for economic status and age, indicated that the relative risk (with 95% C.I.) for cough, and shortness of breath (dyspnoea) was 3.2 (1.6-6.7), and 4.6 (1.2-18.2) respectively.

Childhood acute respiratory infections

Acute lower respiratory infections

Acute respiratory infections (ARIs) are the single most important cause of mortality in children aged less than 5 years, accounting for around 3-5 million deaths annually in this age group(Stansfield, S. et al.1993). Many studies in developing countries have reported on the association between exposure to indoor air pollution and acute lower respiratory infections(Robin, L.F et al 1996; Armstrong,J.R et al.1991). The studies on indoor air pollution from household biomass fuel are reasonably consistent and, as a group, show a significant increase in risk for exposed young children compared with those living in households using cleaner fuels or being otherwise less exposed(Smith, K.et al.2000). Some of the studies carried out in India have reported no association between use of biomass fuels and ARI in children. In a case-control study in children under five years of age in south Kerala(Shah, N.et al.1994), where children with severe pneumonia as ascertained by WHO criteria were compared with those having non severe ARI attending outpatient department, the fuel used for cooking was not a significant risk factor for severe ARI. Non-severe ARI controls may represent the continuum (predecessor) of the cases themselves. Sharma et al in a cross-sectional study in 642 infants dwelling in urban slums of Delhi and using wood and kerosene respectively, did not find a significant difference in the prevalence of acute lower respiratory tract infections and the fuel type.

Upper respiratory tract infections and otitis media

Studies on the relationship between indoor air pollution and acute upper respiratory infections in children both from developed (Samet, J.M.1987,1988) and developing nations(Anderson, H .R 1978,Boleiz, J.S.1989) have not been able to demonstrate the relationship between the two. However, there is strong evidence that exposure to environmental tobacco smoke causes middle ear disease. A recent meta-analysis reported an odds ratio of 1.48 (1.08-2.04) for recurrent otitis media if either parent smoked, and one of 1.38 (1.23-1.55) for middle ear effusion in the same circumstances(Strachan, D.P. and Cook, D.G.1998). A clinic based case-control study of children in rural New York state reported an adjusted odds ratio for otitis media, involving two or more separate episodes, of 1.73 (1.03-2.89) for exposure to wood burning stoves(Daigler, G.E.1991).

Chronic pulmonary diseases
Chronic obstructive pulmonary disease and chronic cor pulmonale

Fig II. Deaths from indoor smoke from solid fuels.(www.WHO.int)

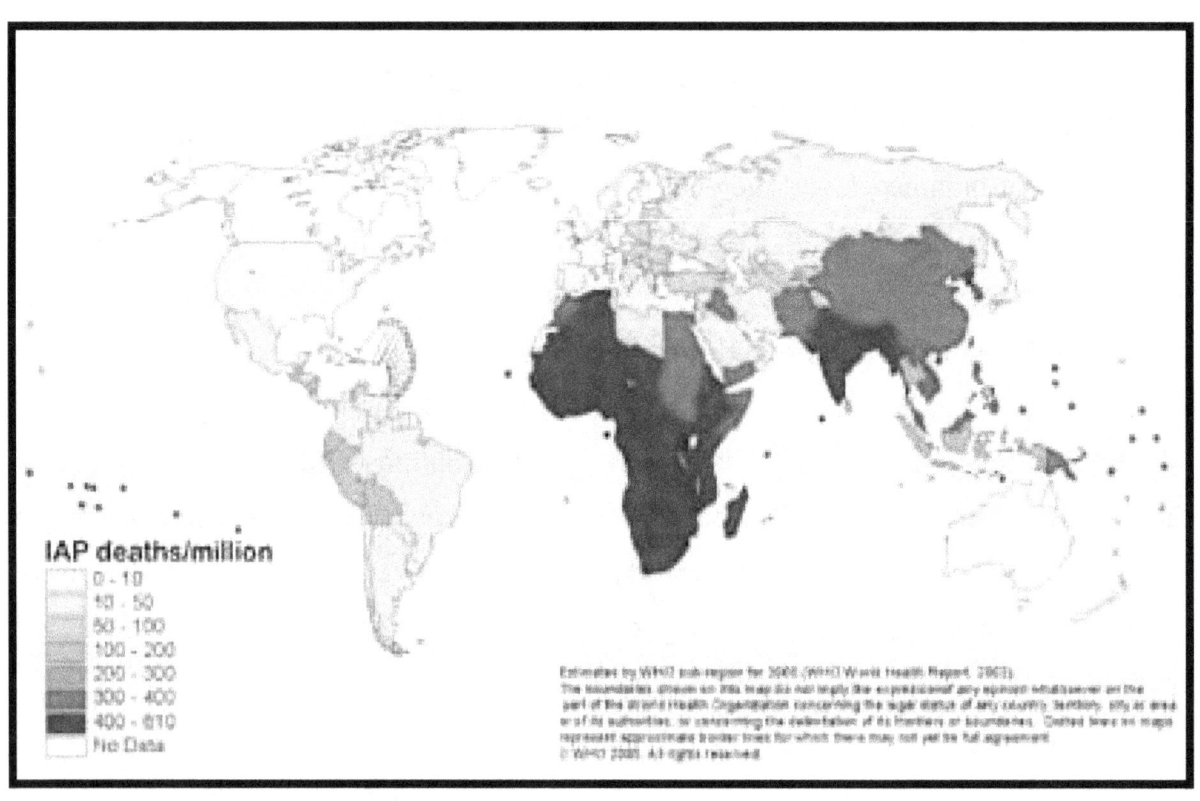

In developed countries, smoking is responsible for over 80% of cases of chronic bronchitis and for most cases of emphysema and chronic obstructive pulmonary disease. Padmavati and colleagues pointed out to the relationship between exposure to indoor air pollutants and chronic obstructive lung disease leading to chronic cor pulmonale. These studies showed that in India, the incidence of chronic cor pulmonale is similar in men and women despite the fact that 75% of the men and only 10% women are smokers. Further analysis of the cases of chronic cor pulmonale in men and women showed that chronic cor pulmonale was more common in younger women. Chronic cor pulmonale seemed to occur 10-15 years earlier in women. The prevalence of chronic cor pulmonale was lower in the southern states than the northern states of India. This is attributed to higher ambient temperatures during most part of the year allowing for greater ventilation in the houses during cooking. The authors attributed this higher prevalence of chronic cor pulmonale in women to domestic air pollution as a result of the burning of solid biomass fuels leading to chronic bronchitis and emphysema which result in chronic cor pulmonale. Subsequent studies in India confirmed these findings(Malik, S.K.1985, Behera, D.1994). Numerous studies from other countries, including ones with cross-sectional and case-control designs, have reported on the association between exposure to biomass smoke and chronic bronchitis or chronic obstructive pulmonary disease(Qureshi, K. Dutt, D.1994, Ellegard, A. 1996,Master, K.M.1974, Pandey, M.R.1984, Norboo, T.1991).

Pneumoconiosis

Pneumoconiosis is a disease of industrial workers occupationally exposed to fine mineral dust particles over a long time. The disease is most frequently seen in miners. Cases of respiratory morbidity who did not respond to routine treatment and whose radiological picture resembled pneumoconiosis have been reported in Ladakh(Norboo, T.1991).However, there are no industries or mines in any part of Ladakh and therefore exposure to dust from these sources was ruled out(Saiyed, H.N.et al.1991,1992). Two factors considered responsible for the development of this respiratory morbidity were (i) Exposure to dust from dust storms. In the spring dust storms occur in many parts of Ladakh. During these storms the affected villages are covered by a thick blanket of fine dust, and the inhabitants are exposed to a considerable amount of dust for several days. The frequency, duration and severity of these dust storms vary considerably from village to village; (ii) Exposure to soot – due to the severe cold in Ladakh, ventilation in the houses is kept at a minimum. The fire place is used for both cooking and heating purposes. To conserve fuel during non-cooking periods, the wood is not allowed to burn quickly but is kept smouldering to prolong its slow heating effect. The inmates are thus exposed to high concentrations of soot. The clinico-radiological investigations of 449 randomly selected villagers from three villages having mild, moderate and severe dust storms showed prevalence of pneumoconiosis of 2.0, 20.1 and 45.3% respectively. The chest radiographs of the villagers showed radiological characteristics which were indistinguishable from those found in miners and industrial workers suffering from pneumoconiosis. The dust concentrations in the kitchens without chimneys varied from 3.22 to 11.30 mg/m3 with a mean of 7.50 mg/m3. The free silica content of these dust samples was below 1%. Dust samples sufficient to allow measurement of the dust concentrations could not be collected during the periods of dust storms. A preliminary analysis of the settled dust samples collected immediately after the storms indicated that about 80% of the dust was respirable and the free silica content ranged between 60 and 70%. Detailed statistical analysis of the data showed that the frequency of dust storms, use of chimney in the houses and age were the most important factors related to the development of pneumoconiosis (Venkaiah, K.1993).Thus, the results of medical and radiological investigations positively established the occurrence of pneumoconiosis in epidemic proportion. Exposure to free silica from dust storms and soot from domestic fuel were suggested as the causes of pneumoconiosis. Low oxygen levels or some other factor associated with high altitude may be an important contributory factor in causation of pneumoconiosis because it has been reported that the miners working at high altitude are more prone to develop pneumoconiosis than their counterparts exposed to the same levels of dust and working in the mines at normal altitude (Odinaev, F.I.1992).

Lung Cancer

The link between lung cancer in Chinese women and\ cooking on an open coal stove has been well established(Smith, K.R. and Liu, Y.1993 Mumford J.L.1995, Xu, Z.et al 1995). Smoking is a major risk factor for lung cancer, however, about two-thirds of the lung cancers were reported in nonsmoking women in China(Gao, Y.T.1996), India(Gupta, R.C.1998) and Mexico(Medina, F.M.1996). The presence of previous lung disease, for example tuberculosis which is common in Indian women, is a risk factor for development of lung cancer in non-smokers(Wu, A.H.1995). The smoke from biomass fuels contain a large number of compounds such as poly aromatic hydrocarbons, formaldehyde, etc. known for their mutagenic and carcinogenic activities, but there is a general lack of epidemiological evidence connecting lung cancer with biomass fuel exposure. The factors associated with rural environment may have a modulating effect on the occurrence of lung cancer and therefore the low incidence of lung cancer in Indian women should not lead to a final conclusion of no link between biomass exposure and lung cancer. It may be concluded that at present there is limited evidence of indoor exposure from coal fires leading to lung cancer and there is no evidence for the biomass fuels. Further investigations are needed to reach definite conclusions.

Pulmonary Tuberculosis

Mishra et al reported the association between use of biomass fuels and pulmonary tuberculosis on the basis of analysis of data collected on 260,000 Indian adults interviewed during the 1992-93 National Family Health Survey. Persons living in households burning biomass fuels were reported to have odd ratio of 2.58 (1.98-3.37) compared to the persons using cleaner fuel, with an adjustment for confounding factors such as separate kitchen, indoor overcrowding, age, gender, urban or rural residence and caste. The analysis further indicated that, among persons aged 20 years and above, 51% of the prevalence of active tuberculosis was attributed to smoke from cooking fuel. However, this study has inherent weakness that the cases of tuberculosis were self reported. There is strong possibility of false reporting as no investigation was done to confirm the reliability of the reporting. Gupta and Mathur(Gupta, B.N. and Mathur, N.1997) have reported similar findings from northern India. This study did not control for the confounding factors except for age. There is experimental evidence to show that the exposure to wood smoke may increases susceptibility of the lungs to infections. Exposure to smoke interferes with the mucociliary defences of the lungs(Houtmeyers, E.et al.1999) and decreases several antibacterial properties of lung macrophages, such as adherence to glass, phagocytic rate and the number of bacteria phagocytosed(Fick, R.B.et al 1984,Beck, B.D.1982). Chronic exposure to tobacco smoke also decreases cellular immunity, antibody production and local bronchial immunity, and there is increased susceptibility to infection and cancer(Johnson, J.D.et al.1990). Indeed, tobacco smoke has been associated with tuberculosis(Altet, M.N.et al 1996,McKenna, M.T.1996). Although the evidence in favour of tuberculosis associated with biomass fuel exposure is extremely weak, there is a theoretical possibility of such an association and considering the public health importance of the problem further experimental and epidemiological studies are necessary.

Cataract

During cooking particularly with biomass fuels, air has to be blown into the fire from time to time especially when the fuel is moist and the fire is smouldering. This causes considerable exposure of the eyes to the emanating smoke. In a hospital-based case-control study in Delhi the use of liquefied petroleum gas was associated with an adjusted odds ratio of 0.62 (0.4-0.98) for cortical, nuclear and mixed, but not posterior sub capsular cataracts in comparison with the use of cow dung and wood (Mohan, M.et al.1989). An analysis of over 170,000 people in India(Mishra, V.K. et al.2000) yielded an adjusted odds ratio for reported partial or complete blindness of 1.32 (1.16-1.50) in respect of persons mainly using biomass fuel compared with other fuels after adjusting for socio-economic, housing and geographical variables; there was a lack of information on smoking, nutritional state, and other factors that might have influenced the prevalence of cataract. It is believed that the toxins from biomass fuel smoke are absorbed systematically and accumulate in the lens resulting in its opacity(Bruce, N.et al.2000). The growing evidence that environmental tobacco smoke causes cataracts is supportive(Shalini et al .1994, West, S.1992).

Adverse Pregnancy Outcome

Low birth weight (LBW) is an important public health problem in developing nations attributed mainly to under nutrition in pregnant women. Low birth weight has serious consequences including increased possibility of death during infancy. Exposure to carbon monoxide from tobacco smoke during pregnancy has been associated with LBW. Levels of carbon monoxide in the houses using biomass fuels are high enough to result in carboxyhaemoglobin levels comparable to those in smokers(Dary, O.et al.1981, Behera, D.et al.1998). In rural Guatemala, babies born to women using wood fuel were 63 g lighter than those born to women using gas and electricity, after adjustment for socio-economic and maternal factors(Boy, E. et al 2000). A study carried out in Ahmedabad reported an excess risk of 50% of stillbirth among women using biomass fuels during pregnancy(Mavlankar, D.V.et al.1991). An association between exposure to ambient air pollution and adverse pregnancy outcome has been widely reported(Dejmek, J.1999, Wang, X.1997, Bobak, M.1999). Considering the association of LBW with a number of disease conditions later in life, there is a need for further studies (ICMR 2001).

References :

A

Ackerman, A S, Toon, O B,Taylor, J P,Johnson, D W, Hobbs, P V, Ferek, R J (1995). Effects of Aerosols on Cloud Albedo : Evaluation of Twomey's Parameterization of Cloud Susceptibility Using Measurements of Ship Tracks. Physics 57: 2684–2695. doi:10.1175/1520-0469(2000)057<2684:EOAOCA>2.0.CO;2.

Ackerman, A S, O B Toon, D E Stevens, A J Heymsfield, V Ramanathan, and E J Welton. (2000). Reduction of tropical cloudiness by soot. Science 288 (5468): 1042-1047. doi:10.1126/science.288.5468.1042. http://www.ncbi.nlm.nih.gov/pubmed/10807573

Agency for Toxic Substances and Disease Registry (July 2006). Toxicological Profile For Hydrogen Sulfide. p. 154. Retrieved 2012-06-20.

Agency for Toxic Substances and Disease Registry(2011). Case Studies in Environmental Medicine Lead (Pb) Toxicity: How are People Exposed to Lead?

Agency for Toxic Substance and Disease Registry. Toxic Substances Portal – Lead.

Agency for Toxic Substances and Disease Registry/Division of Toxicology and Environmental Medicine(2006). ToxFAQs: CABS/Chemical Agent Briefing Sheet: Lead. Archived from the original on 2010-03-04.

Agency for Toxic Substances and Disease Registry (ATSDR)(1995). Toxicological Profile for Asbestos (Draft). U.S. Public Health Service, U.S. Department of Health and Human Services, Atlanta, GA.

Agency for Toxic Substances and Disease Registry (ATSDR)(1997). Toxicological Profile for Cadmium. Draft for Public Comment. Public Health Service, U.S. Department of Health and Human Services, Atlanta, GA..

Agency for Toxic Substances and Disease Registry (ATSDR)(1998). Toxicological Profile for Chromium. U.S. Public Health Service, U.S. Department of Health and Human Services, Atlanta, GA.

Agency for Toxic Substances and Disease Registry (ATSDR)(1999). Toxicological Profile,Chlorinated Dibenzo-p-Dioxins (CDDs).Atlanta, GA: U.S. Public Health Service,U.S. Department of Health and Human Services.

Agency for Toxic Substances and Disease Registry (ATSDR) (December 1990). Public Health Statement, Polycyclic Aromatic Hydrocarbons, December 1990. U.S. Public Health Service, U.S.Department of Health and Human Services,Atlanta, GA.

Air quality and health(2011). www.who.int.

Air pollution causes early deaths(February 21, 2005). BBC.

American Association of Poison Control Centers."AAPCC Annual Data Reports 2007".

American Heart Association (2005). Heart Disease and Stroke Statistics – 2005 Update. http://www.americanheart.org/downloadable/ heart/1105390918119HDSStats2005Update.pdf

American Heart Association (AHA)(June 1, 2010). AHA Scientific Statement Circulation .vol. 121 no. 21 **2331-2378**

Amunom, I, Stephens, L. J., Conklin, D. J., Srivastava, S., Bhatnagar, A, and Prough, R. A.(2005). Several Cytochromes P450 Are Aldehyde Monooxygenases. Weiner, H.Enzymology and Molecular Biology of Carbonyl Metabolism (12), 118-123. Layafette, IN, Purdue Press. Ref Type: Serial

Anderson, H .R(1978). Respiratory abnormalities in Papua New Guinea children. The effects of locality and domestic wood smoke pollution.Int J Epidemiol 7: 63.

Andersen, Z. J., Kristiansen, L. C., Andersen, K. K., Olsen, T. S., Hvidberg, M., Jensen, S. S., Raaschou-Nielsen, O. (2011). Stroke and Long-Term Exposure to Outdoor Air Pollution From Nitrogen Dioxide: A Cohort Study. Stroke; a journal of cerebral circulation. doi:10.1161/STROKEAHA.111.629246.

Andersen, Z. J., Hvidberg, M., Jensen, S. S., Ketzel, M., Loft, S., Sorensen, M., Raaschou-Nielsen, O. (2011). Chronic obstructive pulmonary disease and long-term exposure to traffic-related air pollution: a cohort study. [Research Support, Non-U.S. Gov't]. American journal of respiratory and critical care medicine, 183(4), 455-461. doi:10.1164/rccm.201006-0937OC.

Andersen, Z. J., Bonnelykke, K., Hvidberg, M., Jensen, S. S., Ketzel, M., Loft, S., Raaschou-Nielsen, O. (2011). Long-term exposure to air pollution and asthma hospitalisations in older adults: a cohort study. Thorax. doi:10.1136/thoraxjnl-2011-200711

Anese, M,. Manzocco, L; Calligaris, S; Nicoli, MC (2013). Industrially Applicable Strategies for Mitigating Acrylamide, Furan and 5-Hydroxymethylfurfural in Food. Journal of agricultural and food chemistry: 130528102950009. doi:10.1021/jf305085r. PMID 23627283.

Angier, Natalie (August 21, 2007). "The Pernicious Allure of Lead". New York Times.

Armstrong, J.R. and Campbell, H(1991). Indoor air pollution exposure and lower respiratory infections in young Gambian children.Int J Epidemiol 20: 424.

Atmanand et al. (2009). Energy and Sustainable Development-An Indian Perspective. World Academy of Science.

ATSDR (1998). Toxicological profile for chlorinated dibenzo-p-dioxins (CDDs). Agency for Toxic Substances and Disease Registry.Atlanta, GA.http://www.atsdr.cdc.gov/toxprofiles/tp104.pdf. 197033.

Aylward LL, Goodman JE ,Charnley G, Rhomberg LR (2008). A margin-of-exposure approach to assessment of noncancer risks of dioxins based on human exposure and response data. Environ Health Perspect, 116: 1344-1351 .197068.

Ayres, Robert U. and Ayres, Edward H. (2009). Crossing the Energy Divide: Moving from Fossil Fuel Dependence to a Clean-Energy Future. Wharton School Publishing. p. 36. ISBN 0-13-701544-5.

B

Bakhiya, N., Appel, KE .(2010). Toxicity and carcinogenicity of furan in human diet. Archives of toxicology 84 (7): 563–78. doi:10.1007/s00204-010-0531-y. PMID 20237914.

Baruah BK,D Baruah and M Das (1996 a): Sources and characteristics of Paper mill effluent,.Environment and Ecology. 14 (3) 686 – 689.

Beate Ritz and Michelle Wilhelm (2008).Air Pollution Impacts on Infants and Children.
http://www.environment.ucla.edu/ reportcard / article.asp?parentid=1700.

Beatriz Fátima Alves de Oliveira ,Eliane Ignotti ,Paulo Artaxo ,Paulo Hilário do Nascimento Saldiva .Washington Leite Junger ,Sandra Hacon(2012). Risk assessment of PM2.5 to child residents in Brazilian Amazon region with biofuel production. Environmental Health 2012, 11:64. doi:10.1186/1476-069X-11-64.

Beck, B.D. and Brain, J.D(1982 ,SPU5). Prediction of the pulmonary toxicity of respirable combustion products from residential wood and coal stoves. Proceedings of the Residential Wood and Coal Combustion Special Conference. Air Pollution Control Association,Pittsburg.

Bedard PL, Krzyzanowska MK, Pintilie M, Tannock IF (2007). "Statistical power of negative randomized controlled trials presented at American Society for Clinical Oncology annual meetings". J. Clin. Oncol. 25 (23): 3482–7. doi:10.1200/JCO.2007.11.3670. PMID 17687153

Behera, D., Dash, S. and Malik, S.K.(1998) Blood carboxyhaemoglobin levels following acute exposure to smoke of biomass fuels. Indian J Med Res 88: 522.

Behera, D., Dash, S. and Yadav, S.P.(1991). Blood carboxyhaemoglobin in women exposed to different cooking fuels. Thorax 46: 344.

Behera, D., Jindal, S.K. and Malhotra, H.S.(1994). Ventilatory function in nonsmoking rural Indian women using different cooking fuels. Respiration 61: 89.

Bell ML, Dominici F, Ebisu K, Zeger SL, Samet JM. 2007a. Spatial and temporal variation in PM(2.5) chemical composition in the United States for health effects studies. Environ Health Perspect 115(7):989-995.

Benjamin Stephen Grandey, Philip Stier (2010).Investigating aerosol–cloud interactions. Atmospheric, Oceanic and Planetary Physics,Department of Physics,University of Oxford.

Bergeson, Lynn L. (2008). The proposed lead NAAQS: Is consideration of cost in the clean air act's future?. Environmental Quality Management 18: 79. doi:10.1002/tqem.20197.

Bhatnagar A and Srivastava SK (1992) Aldose reductase: congenial and injurious profiles of an enigmatic enzyme.Biochem.Med.Metab Biol. 48:91-121.

Blumenthal, Ivan (1 June 2001).Carbon monoxide poisoning. J R Soc Med (The Royal Society of Medicine) 94 (6): 270–272. PMC 1281520. PMID 11387414.

Bobak M, Leon DA. 1999. Pregnancy outcomes and outdoor air pollution: an ecological study in districts of the Czech Republic 1986-8. Occup Environ Med 56(8):539-543.

Boleiz, J.S.(1989). Domestic air pollution from biomass burning in Kenya. Atamos Environ 23: 1677.

Boy, E., Bruce, N. and Delgado, H. (2000 ,In Press)Birth weight and exposure to kitchen wood smoke during pregnancy. In: Child and Adolscent Health. World Health Organization, Geneva.

Bradford G. Hill, Oleg Barski, and Aruni Bhatnagar(2006).Getting to the Heart of Pollution: Is Pollution a New Risk Factor for Cardiovascular Disease?.Sustain Fall/ Winter 2006,Issue 13 . Published by the Kentucky Institute for the Environment and Sustainable Development,University of Louisville,203 Patterson Hall,Louisville, Kentucky 40292

Brightwell J, Fouillet X, Cassano-Zoppi A-L, Gatz R, Duchosal F (1986). Neoplastic and functional changes in rodents after chronic inhalation of engine exhaust emissions. In: Ishinishi N, Koizumi A, McClellan RO, Stober W, eds. Carcinogenic and mutagenic effects of diesel engine exhaust. Proceedings of the Symposium on Toxicological Effects of Emissions from Diesel Engines, Tsukuba City, Japan, July 26-28, 1986. New York, NY: Elsevier Science Publishers, pp. 471-487.

Brook, RD; Rajagopalan, S; Pope, CA III; Brook, JR; Bhatnagar, A (2010). Particulate matter air pollution and cardiovascular disease: An update to the scientific statement from the American Heart Association. Circulation 121: 2331–2378.

Bruce, N., Perez-padilla, R. and Albalak, R.(2000) .Indoor air pollution in developing countries: A major environmental and public health challenge. Bull World Health Organ 78: 1078.

C

California Environmental Protection Agency (2005). Technical Support Document for Describing Available Cancer Potency Factors. Available:http://www.oehha.ca.gov/air/hot_spots/pdf/May2005Hotspots.pdf.

Carol Potera (2008). Air Pollution: Salt Mist Is the Right Seasoning for Ozone. Environ Health Perspect 116 (7): A288. PMC 2453175. PMID 18629329.

Cathryn Tonne and Paul Wilkinson (2012). Long-term exposure to air pollution is associated with survival following acute coronary syndrome. European Heart Journal. Published February 19, 2013.doi:10.1093/eurheartj/ehs480.

Centers for Disease Control and Prevention, National Environmental Public Health Tracking Network(accessed 2009), Carbon Monoxide Poisoning.

Central Pollution Control Board (November, 2010). Study of the Exhaust Gases from different fuel based vehicles for Carbonyls and Methane Emissions.

Chang RL, Wood AW, Conney AH, Yagi H, Sayer JM, Thakker DR, Jerina DM, and Levin W (1987) Role of diaxial versus diequatorial hydroxyl groups in the tumorigenic activity of a benzo[a]pyrene bay-region diol epoxide. Proc.Natl.Acad.Sci.U.S.A 84:8633-8636.

Charlson, R.J., S E Schwartz, J M Hales, R D Cess, J A Coakley, J E Hansen, and D J Hofmann (1992). Climate forcing by anthropogenic aerosols. Science 255 (5043): 423–30. doi:10.1126/science.255.5043.423. PMID 17842894.

Chiefs of Ontario, Effects on Aboriginals from the Great Lakes Environment Project(EAGLE). Fact Sheet 11: Dioxins and Furans http://www.chiefs-ofontario.org/eagle/factsheet11.htm

Christopher H. Goss, Stacey A. Newsom, Jonathan S. Schildcrout, Lianne Sheppard and Joel D. Kaufman (2004).Effect of Ambient Air Pollution on Pulmonary Exacerbations and Lung Function in Cystic Fibrosis.

Clifton VL, Giles WB, Smith R, Bisits AT, Hempenstall PA, et al. 2001. Alterations of placental vascular function in asthmatic pregnancies. Am J Respir Crit Care Med 164(4):546-553.

C.Michael Hogan. (2010). Abiotic factor. Encyclopedia of Earth. eds Emily Monosson and C. Cleveland. National Council for Science and the Environment. Washington DC.

Cohen, Alan R.; Trotzky, Margret S.; Pincus, Diane (1981). Reassessment of the Microcytic Anemia of Lead Poisoning. Pediatrics 67 (6): 904–906. PMID 7232054.

Collins, Nick (April 18, 2012). Exhaust fumes are twice as deadly as roads, study claims. The Telegraph.

Committee of the Environmental and Occupational Health Assembly of the American Thoracic Society. (1996). Health effects of outdoor air pollution. [Comparative Study Review]. American journal of respiratory and critical care medicine, 153(1), 3-50.

Committee on Environmental Health (2004). "Ambient Air Pollution: Health Hazards to Children". Pediatrics 114 (6): 1699–1707. doi:10.1542/peds.2004-2166. PMID 15574638.

D

Daigler, G.E., Markello, S.J. and Cummings, K.M. (1991). The effect of indoor air pollutants on otitis media and asthma in children.Laryngoscope 101: 293

Dale Hattis, Ph.D., Clark University (1995).

Daly, A. and P. Zannetti(2007). Ambient Air Pollution . Chapter 1,An Introduction to Air Pollution – Definitions, Classifications, and History. pp 1-14.Publisher The Arab School for Science and Technology (ASST)(http://www.arabschool.org.sy) and The Enviro Comp Institute (http://www.envirocomp.org/).

Danaei G, Vander Hoorn S, Lopez AD, Murray CJ, Ezzati M (2005). Causes of cancer in the world: comparative risk assessment of nine behavioural and environmental risk factors. Lancet 366 (9499): 1784–93. doi:10.1016/S0140-6736(05)67725-2. PMID 16298215

Dangerous Japanese 'Detergent Suicide' Technique Creeps Into U.S. (March 13, 2009). Wired.com (Wired magazine).

Dary, O., Pineda, O. and Belizan, J.(1981).Carbon monoxide in dwellings in poor rural areas of Guatemala. Bull Environ Contam Toxicol. 26: 24.

Davis, Devra (2002). When Smoke Ran Like Water: Tales of Environmental Deception and the Battle Against Pollution. Basic Books. ISBN 0-465-01521-2.

Delin G.N., Essaid H.I., Cozzarelli I.M., Lahvis M.H., and Bekins B.A.,(1998).http://mn.water.usgs.gov/projects/bemidji/results/fact-sheet.pdf.

Della Porta G; Dragani TA; Sozzi G (1987). Carcinogenic effects of infantile and long-term 2,3,7,8-tetrachlorodibenzo-p-dioxin treatment in the mouse. Tumori, 73: 99-107. 197405.

Dockery, D.W., Pope, C.A. 3rd., Xu, X., Spengler, J.D., Ware, J.H., Fay, M.E., et al. (1993). An association between air pollution and mortality in six U.S. cities. N.Engl. J. Med. 329, 1753–1759.

Dominick DalSanto. The Encyclopedia of Dust Collection

Dragan YP; Schrenk D (2000). Animal studies addressing the carcinogenicity of TCDD (or related compounds) with an emphasis on tumour promotion. Food Addit Contam, 17: 289-302. 197243.

Dutt, D.(1996). Effect of indoor air pollution on the respiratory system of women using different fuels for cooking in an urban slum ofPondicherry. Natl Med J India 9: 113.

E

E.A. Davidson & W. Kingerlee (1997). A global inventory of nitric oxide emissions from soils. Nutrient Cycling in Agroecosystems 48: 37–50. doi:10.1023/A:1009738715891.

E.J. Calabrese and E.M. Kenyon. (1991). Air Toxics and Risk Assessment. Lewis Publishers, Chelsea, MI.

Ellegard, A.(1996).Cooking fuel smoke and respiratory symptoms among women in low income areas in Maputo. Environ Health Perspect 104: 980.

Elschenbroich, C.; Salzer, A. (2006). Organometallics: A Concise Introduction (2nd ed.). Weinheim: Wiley-VCH. ISBN 3-527-28165-7.

Emond C; Birnbaum LS; DeVito MJ (2004). Physiologically based pharmacokinetic model for developmental exposures to TCDD in the rat. Toxicol Sci, 80: 115-133. 197315.

Environmental protection agency. How nitrogen oxides affect the way we live and breathe. Archived from the original on 2008-07-16.

Estimated deaths & DALYs attributable to selected environmental risk factors, by WHO Member State (2002).

European Food Safety Authority (2011). EFSA Journal 9 (9): 2347. doi:10.2903/j.efsa.2011.2347.

F

Farrah J. Mateen & Robert D. Brook(March 23 ,2011). Air pollution as an emerging global risk factor for stroke JAMA;305(12):1240-1.

Felicity Barringer (February 18, 2012). Scientists Find New Dangers in Tiny but Pervasive Particles in Air Pollution. The New York Times. Fine atmospheric particles — smaller than one-thirtieth of the diameter of a human hair — were identified more than 20 years ago as the most lethal of the widely dispersed air pollutants in the United States. Linked to both heart and lung disease, they kill an estimated 50,000 Americans each year.

Ferek, R J; Timothy Garrett, P V Hobbs, Scott Strader, Doug Johnson, J P Taylor, Kurt Nielsen, et al. (2000). "Drizzle Suppression in Ship Tracks". Journal of the Atmospheric Sciences 57 (16): 2707–2728. doi:10.1175/1520-0469(2000)057<2707:DSIST>2.0.CO;2.

Forster, Piers; Venkatachalam Ramaswamy, Paulo Artaxo, Terje Berntsen, Richard Betts, David W Fahey, James Haywood, et al. (2007). Contribution of Working Group I to the Fourth Assessment Report of the Intergovernmental Panel on Climate Change In Climate Change 2007: The Physical Science Basis. In S. Solomon, D. Qin, M. Manning, Z. Chen, M. Marquis, K.B. Averyt, M.Tignor, and H.L. Miller. Cambridge, United Kingdom and New York, NY, USA: Cambridge University Press. pp. 129–234.

G

Ganong, William F (2005). Review of medical physiology (22 ed.). McGraw-Hill. p. 684. ISBN 0-07-144040-2. Retrieved May 2009.

Gao, Y.T.(1996) Risk factors for lung cancer among nonsmokers with emphasis on life-style factors. Lung Cancer 14(Suppl.1): S39.

Garshick E, Schenker MB, Munoz A, Seg-Al M, Smith TJ, Woskie SR, Hammond SK, Speizer FE (1987a). A case-control study of lung cancer and diesel exhaust exposure in railroad workers. Am Rev Respir Dis 135(6):1242-1248.

Garshick E, Schenker MB, Munoz A, Segal M, Smith TJ, Woskie SR, Hammond SK, Speizer FE (1988). A retrospective cohort study of lung cancer and diesel exhaust exposure in railroad workers. Am Rev Respir Dis 137:820-825.

Garshick E, Schenker MB, Woskie SR, Speizer FE (1987b). Exposure to asbestos among active railroad workers. Am J Ind Med 12:399-406.

Gehring, U., Wijga, A. H., Brauer, M., Fischer, P., de Jongste, J. C., Kerkhof, M., Brunekreef, B. (2010). Traffic-related air pollution and the development of asthma and allergies during the first 8 years of life. [Research Support, Non-U.S. Gov't]. American journal of respiratory and critical care medicine, 181(6), 596-603. doi:10.1164/rccm.200906-0858OC.

Golub, Mari S., ed. (2005). Metals, fertility, and reproductive toxicity. Boca Raton, Fla.: Taylor and Francis. p. 153. ISBN 978-0-415-70040-5

Grandjean, P. (1978). Widening perspectives of lead toxicity. Environmental Research 17 (2): 303–321. doi:10.1016/0013-9351(78)90033-6. PMID 400972.

Gridlocked Delhi: six years of career lost in traffic jams (September 5, 2010). India Today.

Griffin L.F. and Calder J. A., Applied and environmental microbiology, May 1977, Vol. 33, No. 5, pp. 1092-1096.

Gupta, R.C. (1998). Primary bronchogenic carcinoma: Clinical profile of 279 cases from mid-west Rajasthan. Indian J Chest Dis Allied Sci 40: 109.

Gupta, B.N. and Mathur, N.(1997). A study of the household environmental risk factors pertaining to respiratory disease. Energy
Environ Rev 13: 61 .

H

Hansen, J.; Sato, M.; Ruedy, R. (1997). Radiative forcing and climate response. Journal of Geophysical Research 102 (D6): 6831–6864. Bibcode:1997JGR...102.6831H. doi:10.1029/96JD03436.

Håkan Törnqvist, Nicholas L. Mills, Manuel Gonzalez, Mark R. Miller, Simon D. Robinson, Ian L. Megson, William MacNee, Ken Donaldson, Stefan Söderberg, David E. Newby, Thomas Sandström, and Anders Blomberg(1965).Persistent Endothelial Dysfunction in Humans after Diesel Exhaust Inhalation .

Haywood, James; Boucher, Olivier (2000). Estimates of the direct and indirect radiative forcing due to tropospheric aerosols: A review. Reviews of Geophysics 38 (4): 513. Bibcode:2000RvGeo..38..513H. doi:10.1029/1999RG000078. Retrieved August 11, 2012.

Hemminki K., Niemi M.L. (1982) Int. Arch. Occup. Environ. Health 51 (1): 55-63.

Holder G, Yagi H, Dansette P, Jerina DM, Levin W, Lu AY, and Conney AH (1974). Effects of inducers and epoxide hydrase on the metabolism of benzo(a)pyrene by liver microsomes and a reconstituted system: analysis by high pressure liquid chromatography. Proc.Natl.Acad.Sci.U.S.A 71:4356-4360.

Holland WW, Reid DD. The urban factor in chronic bronchitis. Lancet.445-448.

Holleman, A. F.; Wiberg, E. (2001), Inorganic Chemistry, San Diego: Academic Press, ISBN 0-12-352651-5.

Hong, Y.-C., Lee J.-T., Kim, H., Ha, E.-H., Schwartz, J., and Christiani, D.C.(2002). Effects of Air Pollutants on Acute Stroke Mortality. Environ. Health Perspect.110: 187-191.

Houtmeyers, E., Gosselink, R., Gayan-Ramirez, G. and Decramer, M. (1999). Regulation of mucociliary clearance in health and disease.Eur Respir J 13: 1177.

Hu, Howard (1991). Knowledge of diagnosis and reproductive history among survivors of childhood plumbism. American Journal of Public Health 81 (8): 1070–1072. doi:10.2105/AJPH.81.8.1070. PMC 1405695. PMID 1854006.

Huynh M, Woodruff TJ, Parker JD, Schoendorf KC (2006). Relationships between air pollution
and preterm birth in California. Paediatr Perinat Epidemiol 20(6):454-461.

I

IARC(June 12 ,2012).IARC:Diesel Engine Exhaust Carcinogenic. Press Release N° 213.

ICMR.Indoor Air Pollution In India – A Major Environmental And Public Health Concern (May,2001).ICMR Bulletin.Vol.31,No.5.ISSN 0377-4910

Information for the Community Lead Toxicity. Agency for Toxic Substances and Disease Registry.

International Agency for Research on Cancer(2004) IARC Classifies Formaldehyde as Carcinogenic to Humans. Available: http://www.iarc.fr/ENG/Press_Releases/archives/pr153a.html .

International Energy Agency, France(2011).CO2 Emissions From Fuel Combustion Highlights, 2011 Edition.

Iowa State University Extension (May 2004). "The Science of Smell Part 1: Odor perception and physiological response". PM 1963a.

IPCC (2001).6.7.8 Discussion of Uncertainties. IPCC Third Assessment Report - Climate Change. Retrieved 14 July 2012.

Ishinishi N, Kuwabara N, Nagase S, Suzuki T, Ishiwata S, Kohno T(1986). Long-term inhalation studies on effects of exhaust from heavy and light duty diesel engines on F344 rats. In: Ishinishi N, Koizumi A, McClellan RO, Stober W, eds. Carcinogenic and mutagenic effects of diesel engine exhaust. Proceedings of the Symposium on Toxicological Effects of Emissions from Diesel Engines, Tsukuba City, Japan, July 26-28, 1986. New York, NY: Elsevier Science Publishers, pp. 329-348.

J

Jagadish Prasad, P. (2010). Conceptual Pharmacology. Universities Press. p. 652. ISBN 978-81-7371-679-9. Retrieved 21 June 2012.

Jemal A, Siegel R, Ward E et al. (2008). "Cancer statistics, 2008". CA Cancer J Clin 58 (2): 71–96. doi:10.3322/CA.2007.0010. PMID 18287387

J .L. Mauderly Environ Health Perspect (1994 October). Toxicological and epidemiological evidence for health risks from inhaled engine emissions.nih.gov. 102(Suppl 4): 165–171.

J.N. Galloway, et al. (September 2004). "Nitrogen cycles: past, present, and future". Biogeochemistry 70 (2): 153–226. doi:10.1007/s10533-004-0370-0.

Johnson, J.D., Houchens, D.P., Kluwe, W.M., Craig, D.K. and Fisher, G.L.(1990) Effects of mainstream and environmental tobacco Smoke on the immune system in animals and humans: A review. Crit Rev Toxicol 20: 369.

J. Patrick Mastin(2006).The Contribution of Air Pollution to the Burden of Cardiovascular Disease. Sustain Fall/Winter 2006,Issue 13 . Published by the Kentucky Institute for the Environment and Sustainable Development, University of Louisville,203 Patterson Hall,Louisville, Kentucky 40292

J. Sunyer (2001). "Urban air pollution and Chronic Obstructive Pulmonary disease: a review". European Respiratory Journal 17 (5): 1024–1033. doi:10.1183/09031936.01.17510240. PMID 11488305.

K

Kaatsch, P (June,2010). "Epidemiology of childhood cancer". Cancer treatment reviews 36 (4): 277–85. doi:10.1016/j.ctrv.2010.02.003. PMID 20231056.

Kannan S, Misra DP, Dvonch JT, Krishnakumar A.(2006). Exposures to airborne particulate matter and adverse perinatal outcomes: a biologically plausible mechanistic framework for exploring potential effect modification by nutrition. Environ Health Perspect .114(11):1636-1642.

Kapitulnik J, Levin W, Conney AH, Yagi H, and Jerina DM (1977) Benzo[a]pyrene 7,8-dihydrodiol is more carcinogenic than benzo[a]pyrene in newborn mice.Nature 266:378-380.

Kaufman, Y. J.; Fraser, Robert S. (1997). "The Effect of Smoke Particles on Clouds and Climate Forcing.". Science 277 (5332): 1636–1639. doi:10.1126/science.277.5332.1636..

Kavita Gulati,Basudeb Banerjee,Shyam Bala Lall and Arunabha Ray (July,2010).Effects of diesel exhaust metals and pesticides on various organ systems: Possible mechanisms and strategies for prevention and treatment.Indian Journal of Experimental Biology.Vol.48,pp.710-721.

Keita Ebisu, Michelle L. Bell.(2012). Airborne PM2.5 Chemical Components and Low Birth Weight in the Northeastern and Mid-Atlantic Regions of the United States. ehponline.org. http://dx.doi.org/10.1289/ehp.1104763.

Kinawy A. A., Impact of gasoline inhalation on some neurobehavioural characteristics of male rats, BMC Physiology 2009, 9:21

Kolata, Gina (January 26, 1993). "Carbon Monoxide Gas Is Used by Brain Cells As a Neurotransmitter". The New York Times. Retrieved May 2, 2010.

Koren, Ilan, Yoram J Kaufman, Lorraine a Remer, and Jose V Martins (2004). Measurement of the effect of Amazon smoke on inhibition of cloud formation. Science 303 (5662) (February): 1342-5. doi:10.1126/science.1089424. http://www.ncbi.nlm.nih.gov/pubmed/14988557.

Kurt Straif, MD MPH PhD (IARC)(April , 2013). International Scientific Symposium .Emerging Issues in Environmental and Occupational Health: Mining and Construction in Transition Economies .The IARC Monographs Programme, Vol. 105.The carcinogenicity of diesel engine exhaust, gasoline engine exhaust and some nitroarenes .

L

Laura Perez, Fred Lurmann, John Wilson, Manuel Pastor, Sylvia J. Brandt, Nino Künzli, Rob McConnell (2012). Near-Roadway Pollution and Childhood Asthma:Implications for Developing "Win-Win" Compact Urban Development and Clean Vehicle Strategies. ehponline.org. http://dx.doi.org/10.1289/ehp.1104785.

Laurence, D. R. (1966). Clinical Pharmacology(Third Edition).

Lead Paint Information(Archived on February 2,2012). Master Painters, Australia.

Levin, R.; Brown, M. J.; Kashtock, M. E. et al. (2008). Lead Exposures in U.S. Children, 2008: Implications for Prevention. Environmental Health Perspectives 116 (10): 1285–1293. doi:10.1289/ehp.11241. PMC 2569084. PMID 18941567.

Levin W, Chang RL, Wood AW, Thakker DR, Yagi H, Jerina DM, and Conney AH (1986) Tumorigenicity of optical isomers of the diastereomeric bay-region 3,4-diol-1,2- epoxides of benzo(c)phenanthrene in murine tumor models. Cancer Res. 46:2257-2261.

Lewis, R.J. Sax's Dangerous Properties of Industrial Materials(1996). 9th ed. Volumes 1-3. New York, NY: Van Nostrand Reinhold.

Lide, David R., ed. (2006). CRC Handbook of Chemistry and Physics (87th ed.). Boca Raton, FL: CRC Press. ISBN 0-8493-0487-3.

Li, L; Hsu, A; Moore, PK (2009). Actions and interactions of nitric oxide, carbon monoxide and hydrogen sulphide in the cardiovascular system and in inflammation—a tale of three gases! Pharmacology & therapeutics 123 (3): 386–400. doi:10.1016/j.pharmthera.2009.05.005. PMID 19486912.

Lin CM, Li CY, Yang GY, Mao IF. 2004. Association between maternal exposure to elevated ambient sulfur dioxide during pregnancy and term low birth weight. Environ Res 96(1):41-50.

Liu S, Krewski D, Shi Y, Chen Y, Burnett RT. 2003. Association between gaseous ambient air pollutants and adverse pregnancy outcomes in Vancouver, Canada. Environ Health Perspect 111(14):1773-1778.

Louwies, T; Int Panis, L; Kicinski, M; De Boever, P; Nawrot, Tim S (2013). Retinal Microvascular Responses to Short-Term Changes in Particulate Air Pollution in Healthy Adults. Environmental Health Perspectives. doi:10.1289/ehp.1205721.

M

MacIntosh DL (2000). *Human exposure assessment*. Geneva, World Health Organization (Environmental Health Criteria 214).

Madsen C, Gehring U, Walker SE, Brunekreef B, Stigum H, et al. 2010. Ambient air pollution exposure, residential mobility and term birth weight in Oslo, Norway. Environ Res .110(4):363-371.

Maisonet M, Correa A, Misra D, Jaakkola JJ. 2004. A review of the literature on the effects of ambient air pollution on fetal growth. Environ Res 95(1):106-115.

Mahatnirunkul V., Towprayoon S., Bashkin V.(2002). Application of the EPA hydrocarbon spill screening model to a hydrocarbon contaminated site in Thailand, Land Contamination & Reclamation, 10 (1).

Malik, S.K.(1985). Exposure to domestic cooking fuels and chronic bronchitis. Indian J Chest Dis Allied Sci 27: 171.

Manas Ranjan Ray & Twisha Lahiri (2010). Air Pollution and its Effects on Health – Case Studies, India.Chittaranjan National Cancer Institute,Kolkata.

Marc M. Baum, Eileen S. Kiyomiya, Sasi Kumar, Anastasios M. Lappas, and Harry C. Lord 111'.Remote Sensing Of Criteria And Non-Criteria Pollutants Emitted In The Exhaust Of On-Road Vehicles Department of Chemistry, Oak Crest Institute of Science, 13300 Brooks Dr, Suite B, Baldwin Park, CA 91706, 'Air Instruments & Measurements, Inc, 13300 Brooks Dr, Suite A, Baldwin Park, CA 9 1706.

Marino, P. E., Landrigan, P. J., Graef, J., Nussbaum, A., Bayan, G., Boch, K., Boch, S. (1990). A case report of lead paint poisoning during renovation of a Victorian farmhouse. American Journal of Public Health 80 (10): 1183–1185. doi:10.2105/AJPH.80.10.1183. PMC 1404824. PMID 2119148.

Maroziene L, Grazuleviciene R.(2002). Maternal exposure to low-level air pollution and pregnancy outcomes: a population-based study. Environ Health 1(1):6.

Mary Hardin and Ralph Kahn. Aerosols and Climate Change.

Master, K.M.(1974) .Air pollution in New Guinea. Cause of chronic pulmonary disease among stone-age natives in the highlands. J Am Med Assoc 228: 1635.

Matthew Barth and Kanok Boriboonsomsin (November, 2009). "Real-World CO2 Impacts of Traffic Congestion". Transportation Research Record: Journal of the Transportation Research Board 2058: 163–171. doi:10.3141/2058-20.

Mavlankar, D.V., Trivedi, C.R. and Gray, R.H. (1991).Levels and risk factors for perinatal mortality in Ahmedabad, India. Bull World Health Organ 69: 435.

McCann J, Horn L, Girman J, Nero AV. (1986). Potential Risks from Exposure to Organic Carcinogens in Indoor Air LBL-22474. Berkeley, CA:Lawrence Berkeley Laboratory.

McKenna, M.T. (1996)The association between occupation and tuberculosis.A population-based survey. Am J Respir Crit Care Med 154: 587,.

Medina, F.M. (1996).Primary lung cancer in Mexico City. A report of 1019 cases. Lung Cancer 14: 185.

Menkes D. B. and Fawcett J. P.(March 1997).Too Easily Lead? Health Effects of Gasoline Additives,Environmental Health Perspectives, Vol. 105, No. 3.

Michael Kymisis, Konstantinos Hadjistavrou (2008). "Short-Term Effects Of Air Pollution Levels On Pulmonary Function Of Young Adults". The Internet Journal of Pulmonary Medicine 9 (2).

Michael T. Kleinman (2000).The Health Effects of Air Pollution on Children. Department of Community and Environmental Medicine.University of California, Irvine. http://www.aqmd.gov/forstudents/health_effects_on_children.html

Mishra, V.K., Retherford, R.D. and Smith K.R.(1999) Biomass cooking fuels and prevalence of tuberculosis in India. Int J Infect Dis 3: 119.

Miller K. A., Siscovick D. S., Sheppard L., Shepherd K., Sullivan J. H., Anderson G. L., Kaufman J. D. (2007). Long-term exposure to air pollution and incidence of cardiovascular events in women . The New England journal of medicine (Research Support, N.I.H., ExtramuralResearch Support, U.S. Gov't, Non-P.H.S.) 356 (5): 447–458. doi:10.1056/NEJMoa054409.

Milton R. Beychok (March 1973). "NOX emission from fuel combustion controlled". The Oil and Gas Journal: 53–56.

Mohan, M., Sperduto, R.D. and Angra, S.K.(1989) India-US case control study of age-related cataracts. Arch Ophthalmol 107: 670.

Moorthy B, Miller KP, Jiang W, and Ramos KS (2002) The atherogen 3-methylcholanthrene induces multiple DNA adducts in mouse aortic smooth muscle cells: role of cytochrome P4501B1. Cardiovasc.Res. 53:1002-1009.

Mumford J.L. (1995).Human exposure and dosimetry of polycyclic aromatic hydrocarbons in urine from Xuan Wei, China with high lung cancer mortality associated with exposure to unvented coal smoke. Carcinogenesis 16: 3031. Mutagenesis (November 2005,).3-Nitrobenzanthrone, a potential human cancer hazard in diesel exhaust and urban air pollution. pp 399-410.

Morello-Frosch R, Jesdale BM, Sadd JL, Pastor M. 2010. Ambient air pollution exposure and full-term birth weight in California. Environ Health 9:44.C.

Miranda M. Loh, Jonathan I. Levy, John D. Spengler, E. Andres Houseman, and Deborah H. Bennett (August 2007).Ranking Cancer Risks of Organic Hazardous Air Pollutants in the United States. Environmental Health Perspectives • Vol. 115 .No. 8 .

Mauderly JL, Jones RK, Griffith WC, Henderson RF, McClellan RO (1987). Diesel exhaust is a pulmonary carcinogen in rats exposed chronically by inhalation. Fund Appl Toxicol 9:208-221.

Moro, S; Chipman, JK; Wegener, JW; Hamberger, C; Dekant, W; Mally, A (2012). Furan in heat-treated foods: Formation, exposure, toxicity, and aspects of risk assessment. Molecular nutrition & food research 56 (8): 1197–211. doi:10.1002/mnfr.201200093. PMID 22641279.

N

Nandini Kumar and Anita Dahiya (2011).Current status of air pollution in India and gaps in the research .Department of Natural Resources ,TERI University .

National Academies(1977). Ozone and other photochemical oxidants p. 23. ISBN 0-309-02531-1.

National Ambient Air Quality Status & Trends in India-2010(2012). Central Pollution Control Board, Ministry Of Environment & Forests.

National Institute of Occupational Health, Ahmedabad (1982) Domestic source of air pollution and its effects on respiratory
system of house wives in Ahmedabad. (Annual Report). p.32.

Needleman, Herbert L.; Schell, Alan; Bellinger, David; Leviton, Alan; Allred, Elizabeth N. (1990). The long-term effects of exposure to low doses of lead in childhood. An 11-year follow-up report. New England Journal of Medicine 322 (2): 83–88. doi:10.1056/NEJM199001113220203. PMID 2294437.

NIOSH Pocket Guide to Chemical Hazards.

3-Nitrobenzanthrone, a potential human cancer hazard in diesel exhaust and urban air pollution: a review of the evidence(November, 2005). Oxford Journals, Life Sciences & Medicine ,Mutagenesis .Vol 20, Issue 6.pp. **399-410.**

NOAA Study Shows Nitrous Oxide Now Top Ozone-Depleting Emission,(August 27, 2009) .NOAA.

Nonequilibrium atmospheric secondary organic aerosol formation and growth. . Proceedings of the National Academy of Sciences of the United States of America. Published online before print ,. doi:10.1073/pnas.1119909109

Norboo, T., Angchuk, P.T. and Yahya, M. (1991)Silicosis in a Himalayan village population: Role of environmental dust. Thorax46: 341.

Norboo, T. (1991). Domestic pollution and respiratory illness in a Himalayan village. Int J Epidemiol 20: 749.

O

Obidike I.R., Maduabuchi I. U., Olumuyiwa S.S.V.(2007) .Testicular morphology and cauda epididymal sperm reserves of male rats exposed to Nigerian Qua Iboe Brent crude oil, J. Vet. Sci, 8(1), 1–5.

Odinaev, F.I.(1992) The characteristics of development and course of pneumoconiosis under the conditions of a mountain climate. Grig Tr Prof Zabol 7: 13, (Russian).

Odinaev, F.I.(1992). The mechanism of the formation of pneumoconiosis under high altitude conditions. Grig Tr Prof Zabol 7: 11,(Russian).

Omaye ST (2002). Metabolic modulation of carbon monoxide toxicity. Toxicology 180 (2): 139–150. doi:10.1016/S0300-483X(02)00387-6. PMID 12324190.

O.P Singh,Air Pollution:Types,sources and Abatements.

OSHA(Retrieved May 2009).OSHA CO guidlines.

P

Padmavati, S. and Joshi, B.(1964) Incidence and etiology of chronic cor pulmonale in Delhi: A necropsy study. Dis Chest 46: 457.

Padmavati, S. and Pathak, S.N. (1959) Chronic cor pulmonale in Delhi.Circulation 20: 343.

Pandey, M.R.(1984). Domestic smoke pollution and chronic bronchitis in a rural community of the hill region of Nepal. Thorax 39:337.

Pandey, M.R. (1984).Prevalence of chronic bronchitis in rural community of the hill region of Nepal. Thorax 39: 331.

Pope CA, Burnett RT, Thun MJ, Calle EE, Krewski D, Ito K, Thurston GD (2002). Lung Cancer, Cardiopulmonary Mortality, and Long-Term Exposure to Fine Particulate Air Pollution.JAMA.1132-1141.

Pope CA III, Burnett RT, Thurston GD, Thun MJ, Calle EE, Krewski D, Godleski JJ.(2004) .Cardiovascular Mortality and Year-round Exposure to Particulate Air Pollution: epidemiological evidence of general pathophysiological pathways of disease. Circulation.;109:71-77.

Pope, C.A. 3rd., Thun, M.J., Namboodiri, M.M., Dockery, D.W., Evans, J.S., Speizer, F.E., et al. (1995). Particulate air pollution as a predictor of mortality in a prospective study of U.S. adults. Am. J. Respir. Crit. Care Med. 151,669–674.

Prockop LD, Chichkova RI (2007).Carbon monoxide intoxication: an updated review. J Neurol Sci 262 (1–2): 122–130. doi:10.1016/j.jns.2007.06.037. PMID 17720201.

Q

Oudot J.(1990). Selective Migration of Low and Medium Molecular Weight Hydrocarbons in Petroleum-ContaminatedTerrestrial Environments, Oil & Chemical Pollution 6, 251-261.

Qureshi, K.(1994) Domestic smoke pollution and prevalence of chronic bronchitis/asthma in a rural area of Kashmir. Indian J Chest Dis Allied Sci 36: 61.

R

Raaschou-Nielsen, O., Andersen, Z. J., Hvidberg, M., Jensen, S. S., Ketzel, M., Sorensen, M., Tjonneland, A. (2011). Air pollution from traffic and cancer incidence: a Danish cohort study. [Research Support, Non-U.S. Gov't]. Environmental health : a global access science source, 10, 67. doi:10.1186/1476-069X-10-67.

Raaschou-Nielsen, O., Andersen, Z. J., Hvidberg, M., Jensen, S. S., Ketzel, M., Sorensen, M., Tjonneland, A. (2011). Lung cancer incidence and long-term exposure to air pollution from traffic. [Research Support, Non-U.S. Gov't]. Environmental health perspectives, 119(6), 860-865. doi:10.1289/ehp.1002353.

Ramesh A, Walker SA, Hood DB, Guillen MD, Schneider K, Weyand EH.(2004). Bioavailability and risk assessment of orally ingested polycyclic aromatic hydrocarbons. Int J Toxicol 23(5):301–333; doi:10.1093/toxsci/kfj016.

Ramos KS, Zhang Y, Sadhu DN, and Chapkin RS (1996).The induction of proliferative vascular smooth muscle cell phenotypes by benzo(a)pyrene is characterized by up-regulation of inositol phospholipid metabolism and c-Ha-ras gene expression. Arch.Biochem.Biophys.332:213-222.

Reeves, Claire .E.; Penkett, Stuart A.; Bauguitte, Stephane; Law, Kathy S.; Evans, Mathew J.; Bandy, Brian J.; Monks, Paul S.; Edwards, Gavin D. et al. (2002). Potential for photochemical ozone formation in the troposphere over the North Atlantic as derived from aircraft observationsduring ACSOE. Journal of Geophysical Research 107 (D23): 4707. doi:10.1029/2002JD002415.

Richard B. Schlesinger(1995).Department of Environmental Medicine, New York University Medical Center, New York, New York, USA .Toxicological Evidence for Health Effects from Inhaled Particulate Pollution: Does it Support the Human Experience?Inhalation Toxicology, Vol. 7, No. 1 , pp 99-109.

Ritz B, Wilhelm M, Hoggatt KJ, Ghosh JK(2007). Ambient air pollution and preterm birth in the environment and pregnancy outcomes study at the University of California, Los Angeles.Am J Epidemiol 166(9):1045-1052.

Ritz B, Yu F(1999). The effect of ambient carbon monoxide on low birth weight among children born in southern California between 1989 and 1993. Environ Health Perspect 107(1):17-25.

Robin, L.F. Less, P.S.J. Winger, M., Steinhoff, M., Moulten, L.H., Santoshom, M. and Correa, A.(1996). Wood-burning stoves and lower respiratory illness in Navajo children. Paediatr Infect Dis J 15: 859.

Rosenfeld, D (1999). TRMM observed first direct evidence of smoke from forest fires inhibiting rainfall. Geophysical Research Letters 26 (20): 3105–3108. Bibcode:1999GeoRL..26.3105R. doi:10.1029/1999GL006066.

Roth D., Herkner H., Schreiber W., Hubmann N., Gamper G., Laggner A.N., Havel C.(July, 2011) . Accuracy of Noninvasive Multiwave Pulse Oximetry Compared With Carboxyhemoglobin From Blood Gas Analysis in Unselected Emergency Department Patients. Ann Emerg Med ,58(1):74–9.

Russell A. Prough, Immaculate Amunom,and Daniel J. Conklin(2006).The Role of Metabolism in Protection Against Cardiotoxic Compounds.Sustain Fall/ Winter 2006 . Published by the Kentucky Institute for the Environment and Sustainable Development,University of Louisville,203 Patterson Hall,Louisville, Kentucky 40292.

S

Saiyed, H.N., Sharma, Y.K., Norboo, T., Sadhu, H.G., Majumdar, P.K. and Kashyap, S.K.(1992). Clincio-radiological and PFT profile in non-occupational pneumoconiosis. Indian J Industr Med 38: 148.

Salam MT, Millstein J, Li YF, Lurmann FW, Margolis HG, et al.(2005). Birth outcomes and prenatal exposure to ozone, carbon monoxide, and particulate matter: results from the Children's Health Study. Environ Health Perspect 113(11):1638-1644.

Samet, J.M., Marbury, M.C. and Spengler, J.D.(1987) .Health effects and sources of indoor air pollution. Part I. Am Rev Respir Dis136: 1486.

Samet, J.M., Marbury, M.C. and Spengler, J.D.(1988). Health effects and sources of indoor pollution (state of the art). Am Rev Respir Dis 137: 221.

Sapkota A, Chelikowsky AP, Nachman KE, Cohen AJ, Ritz B. (2010). Exposure to particulate matter and adverse birth outcomes: a comprehensive review and meta-analysis. Air Quality, Atmosphere & Health:10.1007/s11869-11010-10106-11863.

Sax SN, Bennett DH, Chillrud SN, Kinney PL, Spengler JD (2004).Differences in source emission rates of volatile organic compounds in inner-city residences of New York City and Los Angeles. J Expo Anal Environ Epidemiol 14(suppl 1):S95–109; doi:10.1038/sj.jea.7500364.

Scientific and Technical Advisory Panel(May.2004). the use of bioindicators, biomarkers and analytical methods for the analysis of POPs in developing countries.

Schoeters, Greet; Den Hond, Elly; Dhooge, Willem; Van Larebeke, Nik; Leijs, Marike (2008). Endocrine Disruptors and Abnormalities of Pubertal Development. Basic & Clinical Pharmacology & Toxicology 102 (2): 168–175. doi:10.1111/j.1742-7843.2007.00180.x. PMID 18226071.

Seinfeld, John; Spyros Pandis (1998). Atmospheric Chemistry and Physics: From Air Pollution to Climate Change (2nd ed.). Hoboken, New Jersey: John Wiley & Sons, Inc. p. 97. ISBN 0-471-17816-0.

Selkirk JK, Croy RG, Roller PP, and Gelboin HV (1974).High-pressure liquid chromatographic analysis of benzo(alpha)pyrene metabolism and covalent binding and the mechanism of action of 7,8-benzoflavone and 1,2-epoxy-3,3,3-trichloropropane. Cancer Res. 34:3474-3480.

Seo JH, Leem JH, Ha EH, Kim OJ, Kim BM, et al.(2010). Population-attributable risk of low birthweight related to PM10 pollution in seven Korean cities. Paediatr Perinat Epidemiol 24(2):140-148.

Shah, N., Ramakutty, V., Premila, P.G. and Sathy, N.(1994). Risk factors for severe pneumonia in children in south Kerala: A hospital based case-control study. J Trop Paediatr 40: 201.

Shah PS, Balkhair T, Knowledge Synthesis Group on Determinants of Preterm/LBW Births (2011). Air pollution and birth outcomes: a systematic review. Environ Int 37 (2): 498–516. doi:10.1016/j.envint.2010.10.009. PMID 21112090.

Shalini, V.K. Lothra, M., and Srinivas, I.(1994) Oxidative damage to the eye lens caused by cigarette smoke and fuel wood condensates.Indian J Biochem Biophys 31: 262.

Sharma, S., Sethi, G.R., Rohtagi, A., Chaudhary, A., Shankar, R., Bapna, J.S., Joshi, V. and Sapir, D.G. (1998).Indoor air quality and acute lower respiratory infection in Indian urban slums. Environ Health Perspect 106: 291.

Shriver, Atkins.(2010) Inorganic Chemistry, Fifth Edition. W. H. Freeman and Company; New York, pp. 414.

Singh P. , DeMarini D. M., Dick C. A. J., Tabor D. G., Ryan J. V., Linak W. P., Kobayashi T. and Gilmour M. I.(June 2004).Sample Characterization of Automobile and Forklift Diesel Exhaust Particles and Comparative Pulmonary Toxicity in Mice, Environmental Health Perspectives, Vol- 112, No. 8.

Simi Chakrabarti. 20th anniversary of world's worst industrial disaster. Australian Broadcasting Corporation.

Smith, Donald R.; Flegal, A. Russell. Lead in the Biosphere: Recent Trends. JSTOR 4314280.

Smith, K.R. and Liu, Y.(1993) Indoor air pollution in developing countries In: Epidemiology of Lung Cancer. Lung Biology in Health andDisease. Ed. J Samet. Marcel Dekker, New York, p.151.

Smith, K., Samet, J.M., Romieu, I. and Bruce, N.(2000). Indoor air pollution in developing countries and acute respiratory infections in children. Thorax 55: 518.

Spence family deaths: Inquest told it was NI's worst farming tragedy for 20 years(29 January 2013). BBC News.

Stansfield, S. and Shepherd, D.(1993). Acute respiratory infection. In:Disease Control Priorities in Developing Countries. Eds. D.Jameson, W. Mosle, A. Mesham and J. Bobadilla. Oxford University Press, Oxford, p.67.

Strachan, D.P. and Cook, D.G.(1998). Parental smoke, middle ear disease and adenotonsillectomy in children. Thorax 53: 50.

Study links traffic pollution to thousands of deaths(April 15 ,2008). The Guardian (London, UK: Guardian Media Group). Archived from the original on 20 April 2008.

Sumanth D. Prabhu(2006). Environmental Pollution: Relationship to Cardiac Dysfunction and Heart Disease.Sustain Fall/ Winter 2006 . Published by the Kentucky Institute for the Environment and Sustainable Development, University of Louisville,203 Patterson Hall,Louisville, Kentucky 40292.

Sun, Q., Wang, A., Jin, X., Natanzon A., Duquaine, D., Brook, R.D., Aguinaldo, R.D., Fayad, Z.A., Fuster, V., Lippmann, M., Chen, L.C., and Rajagopalan, S. (2005).Long-term Air Pollution Exposure and Acceleration of Atherosclerosis and Vascular Inflammation in an Animal Model. JAMA 294, 3003-3010.

T

Tancrede M, Wilson R, Zeise L, Crouch EAC (1987). The carcinogenic risk of some organic vapors indoors: a theoretical survey. Atmos Environ 21(10):2187–2205; doi:10.1016/0004-6981(87)90351-9.

The World Bank (2010). Mongolia: Air Pollution in Ulaanbaatar - Initial Assessment of Current Situations and Effects of Abatement Measures.

The World Bank. (2002). Urban Air Pollution, Catching gasoline ad diesel adulteration .

Tingting L and woods J S (2009).Cloning,expression, and biochemical properties of CPOX4 : A genetic variant of coproporphyrinogen oxidase that affects susceptibility to mercury toxicity in humans.Toxicol Sci.109 228.

Tikuisis, P; Kane, DM; McLellan, TM; Buick, F; Fairburn, SM (1992). Rate of formation of carboxyhemoglobin in exercising humans exposed to carbon monoxide. Journal of Applied Physiology 72 (4): 1311–9. PMID 1592720

Tracking Carbon Monoxide . Environmental Public Health Tracking – Florida Dept. of Health.

T.S. Ashton (1948). The Industrial Revolution, 1760-1830, London: Oxford University Press.

Tucker Blackburn, Susan (2007). Maternal, fetal, & neonatal physiology: a clinical perspective. Elsevier Health Sciences. pp. 325. ISBN 1-4160-2944-3.

Twomey, S. (1977). "The influence of pollution on the shortwave albedo of clouds". Journal of the Atmospheric Sciences 34 (7): 1149–1152. doi:10.1175/1520-0469(1977)034<1149:TIOPOT>2.0.CO;2

U

UNEP(2001).India; State of the Environment 2001.Air Pollution .Published by UNEP; pp97-113.

UNEP (2001).Primary and Secondary Sources of Aerosols: Soil dust. Climate Change 2001: Working Group 1.

UNEP (2001).Primary and Secondary Sources of Aerosols: Sea salt. Climate Change 2001: Working Group 1.

UNEP (2001).Primary and Secondary Sources of Aerosols: Primary biogenic aerosols. Climate Change 2001: Working Group 1.

UNEP (2001).Primary and Secondary Sources of Aerosols: Carbonaceous aerosols . Climate Change 2001: Working Group 1.

U.S Consumer Product Safety Commission (accessed 2009-12-04). Carbon Monoxide Questions and Answers.

United States Environmental Protection Agency.Sulfur Dioxide.

US Department of Energy, Lawrence Berkeley National Laboratory (2013). How Can Air Pollution Hurt My Health? . Health Effects of Air Pollution.

U.S. Department of Health and Human Services(1993).Hazardous Substances Data Bank (HSDB, online database). National Toxicology Information Program, National Library of Medicine, Bethesda, MD.

U.S. Department of Health and Human Services (2001). Hazardous Substances Data Bank (HSDB, online database).National Library of Medicine Bethesda, MD.

U.S. Department of Health and Human Services(1993).Registry of Toxic Effects of Chemical Substances (RTECS, online database). National Toxicology Information Program, National Library of Medicine, Bethesda, MD.

U.S. Environmental Protection Agency (1994). Deposition of Air Pollutants to the Great Waters. First Report to Congress. EPA-453/R-93-055. Office of Air Quality Planning and Standards, Research Triangle Park, NC.

U.S. EPA (2010). EPA's Reanalysis of Key Issues Related to Dioxin Toxicity and Response to NAS Comments.

U.S. Environmental Protection Agency (1999). Integrated Risk Information System (IRIS) on Chromium VI. National Center for Environmental Assessment, Office of Research and Development, Washington, DC.

U.S. EPA (U.S. Environmental Protection Agency)(2005).Integrated Risk Information System. Available: http://www.epa.gov/iris/ .

U. S. Environmental Protection Agency.Health and Environmental Impacts of NOx.

U.S.EPA (1980). Health and Environmental Effects Profile for Hydrogen Sulfide p.118-8 ECAO-CIN-026A

U.S. EPA. Priority PBTs : Dioxins and Furans Fact Sheet. Washington, D.C.: Office of Pollution Prevention and Toxics.

U.S. Food and Drug Administration.(Retrieved 2010).CPG Sec. 545.450 Pottery (Ceramics); Import and Domestic – Lead Contamination.

U.S.NIOSH (United States National Institute for Occupational Safety and Health)(April 2007) .

U.S. NIOSH Adult Blood Lead Epidemiology and Surveillance .

U.S.NIOSH(August 1988).Carcinogenic Effects of Exposure to Diesel Exhaust.NIOSH Publications and Products.DHHS (NIOSH) Publication Number 88-116

V

Venkaiah, K., Saiyed, H.N., Sharma, Y.K., Sadhu, H.G. and Kashyap, S.K.(1993) .Multiple logistic model to assess the non-occupational pneumoconiosis risk. Indian J Occup Health 36: 103.

Verma, A; Hirsch, D.; Glatt, C.; Ronnett, G.; Snyder, S. (1993). Carbon monoxide: A putative neural messenger. Science 259 (5093): 381–4. doi:10.1126/science.7678352. PMID 7678352.

W

Waizenegger, J; Winkler, G; Kuballa, T; Ruge, W; Kersting, M; Alexy, U; Lachenmeier, DW (2012). Analysis and risk assessment of furan in coffee products targeted to adolescents. Food additives & contaminants. Part A, Chemistry, analysis, control, exposure & risk assessment 29 (1): 19–28. doi:10.1080/19440049.2011.617012. PMID 22035212.

WECF(Women in Europe for a Common Future). Dangerous Health Effects of Home Burning of Plastics and Waste.Fact sheet. WECF,Blumenstrasse 28 D – 80311 Munich, Germany .wecf@wecf.org.www.wecf.org

Weinstock, B.; Niki, H. (1972). Carbon Monoxide Balance in Nature. Science 176 (4032): 290–2. Bibcode:1972Sci...176..290W. doi:10.1126/science.176.4032.290. PMID 5019781.

Wei Y, Han IK, Shao M, Hu M, Zhang OJ, et al. 2009. PM2.5 constituents and oxidative DNA damage in humans. Environ Sci Technol 43(13):4757-4762.

West, S. (1992). Does smoke get in your eyes? J Am Med Assoc 268:1025.

Welch F., Murray V. S. G., Robins A. G., Derwent R. G., Ryall D. B., Williams M. L., Elliott A. J., Analysis of a petrol plume over England: 18–19 January 1997, Occup Environ Med 1999;56:649–656.

William M. Pierce, Jr, Jian Cai, Ned Smith, Delano Turner, Kevin G. Taylor, and Jason R. Neale. Not Magic, but Mass Spectrometry to Monitor Interactions of Environmental Pollutants and Cardiovascular Components. Sustain Fall/Winter 2006 . Published by the Kentucky Institute for the Environment and Sustainable Development, University of Louisville,203 Patterson Hall,Louisville, Kentucky 40292.

Woodruff TJ, Caldwell J, Cogliano VJ, Axelrad DA.(2000).Estimating cancer risk from outdoor concentrations of hazardous air pollutants in 1990. Environ Res 82(3):194–206;doi:10.1006/enrs.1999.4021.

World Health Organization(WHO) (February 2006). "Cancer".

World Health Organization(1988). Chromium. Environmental Health Criteria 61. Geneva, Switzerland.

World Health Organization (2009).WHO Disease and injury country estimates. http://www.who.int/healthinfo/global_burden_disease/estimates_country/en/index.html.

Worldwatch Institute (2013).Coal, China, and India: A Deadly Combination for Air Pollution? © 2013 Worldwatch Institute | worldwatch@worldwatch.org 1400 16th St. NW, Ste. 430, Washington, DC 20036 |(202) 745-8092.

Wu, L; Wang, R (December 2005). "Carbon Monoxide: Endogenous Production, Physiological Functions, and Pharmacological Applications". Pharmacol Rev 57 (4): 585–630. doi:10.1124/pr.57.4.3. PMID 16382109. Retrieved May 26, 2009.

X

Xu, Z., Kjellstorm, T., Xu, X., Lin, U. and Daqlan, Y. (1995)Air Pollution and Its Effects in China. Ed. B. Chen. World Health Organisation,Geneva, pp.47.

Y

Yale University(2012).Data Explorer :: Indicator Profiles - Environmental Performance Index.

Z

Zenz, C., O.B. Dickerson, E.P. Horvath(1994).Occupational Medicine. 3rd ed. St. Louis, MO.pp.886

Zoidis, John D. (1999). The Impact of Air Pollution on COPD. RT: for Decision Makers in Respiratory Care.

Other sources –

Wikipedia.com.

Caspharma.com.

www.epa.gov.

www.nrdc.org .

www.Cseindia.org.

http://cdn.intechopen.com/pdfs/37042/InTechHydrocarbon_pollution_effects_on_living_organisms_remediation_of_contaminated_environments_and_effects_of_heavy_metals_co_contamination_on_bioremediation.pdf

http://www.toronto.ca/health/pdf/de_technical_appendix.pdf

http://www.nrdc.org/air/transportation/ebd/ebdinx.asp

http://www.rrcap.ait.asia/male/manual/national/02chapter2.pdf

http://www.epa.qld.gov.au/environmental_management/air/air_quality_monitoring/air_pollutants/ozone/

http://www.pbs.org/now/science/smog.html

http://en.wikipedia.org/wiki/Acid_rain

http://www.epa.gov/ttn/atw/pollsour.html http://www.epa.gov/ttn/atw/orig189.html

http://www.epa.gov/air/airtrends/aqtrnd03/

http://epa.gov/radtown/air.htm

http://airquality.charmeck.org

http://en.wikipedia.org/wiki/Air_pollution

http://www.epa.gov/air/basic.html

http://yosemite.epa.gov/R10/AIRPAGE.NSF

http://edugreen.teri.res.in/explore/air/indoor.htm

http://www.epa.gov/iaq/ia-intro.html

http://www.epa.gov/ttn/atw/hlthef/asbestos.html

http://www.epa.gov/ttn/atw/hlthef/cadmium.html

http://www.epa.gov/ttn/atw/hlthef/chromium.html

www.catf.us/goto/dieselhealth

www.catf.us/projects/diesel/videos

www.catf.us/goto/noescape

http://www.ejnet.org/dioxin/

http://www.firerescue1.com/fire-products/hazmat-equipment/articles/968922-The-chemical-suicide-phenomenon/

http://en.wikipedia.org/wiki/Polychlorinated_dibenzodioxins

http://www.reuters.com/article/2007/10/18/us-dioxins-idUSN1836384520071018?feedType=RSS&feedName=healthNews

http://www.stadtentwicklung.berlin.de/umwelt/umweltatlas/ed309_01.htm

http://www.worldcoal.org/coal/uses-of-coal/coal-electricity/

http://thediplomat.com/asia-life/2013/07/killer-smog-east-asia-india-southeast-asia-top-global-air-pollution/

http://info.publicintelligence.net/LARTTAChydrogensulfide.pdf ,

http://info.publicintelligence.net/MAchemicalsuicide.pdf ,

http://info.publicintelligence.net/illinoisH2Ssuicide.pdf ,

http://www.epa.gov/mercury/about.htm

http://www.epa.gov/iris/subst/0692.htm

http://www.epa.gov/mercury/effects.htm

http://www.epa.gov/mercury/eco.htm

http://www.epa.gov/air/toxicair/newtoxics.html

http://www.epa.gov/ttn/atw/hlthef/hapintro.html#5a

http://www.ilo.org/oshenc/part-vii/environmental-health-hazards/item/497-industrial-pollution-in-developing-countries

http://marketspace.thinktosustain.com/2011/10/most-polluted-industrial-clusters-of-india-a-review/#.UfDEJ2Tn_Dc

http://marketspace.thinktosustain.com/2012/07/top-10-carbon-emitting-cities-in-india/#.UfDM_GTn_Dc

http://articles.timesofindia.indiatimes.com/2012-06-06/ranchi/32078463_1_pollution-level-air-pollution-pollution-control

http://indrus.in/society/2013/06/20/polluting_baikal_pulp_and_paper_mill_to_be_shut_26271.html

http://www.clarion.ind.in/index.php/clarion/article/view/37/53

http://www.vri-online.org.uk/apk/pollution-campaign.php

http://web.mit.edu/12.000/www/m2015/2015/solutions_for_industrial_pollution.html

http://www.indiancancersociety.org/pdfs/mumbai-rep2006.pdf

http://medind.nic.in/haa/t05/i1/haat05i1p29.pdf

http://www.who.int/mediacentre/factsheets/fs297/en/.

http://www.who.int/ceh/risks/cehair/en/index.html

http://informahealthcare.com/doi/abs/10.3109/08958379509014274

http://www.epa.gov/ttn/naaqs/standards/pm/s_pm_cr_cd.html

http://www.lakestay.co.uk/whitehavenmininghistory.html

http://www.epa.qld.gov.au/environmental_management/air/air_quality_monitoring/air_pollutants/ozone/

MoreThanOrganic.com. Sulphites in wine.

© Phys.org™ 2003-2013

pharmaceutical – Britannica Online Encyclopedia. Britannica.com.

The Food Standards Agency website.Current EU approved additives and their E Numbers.

www.ingramcontent.com/pod-product-compliance
Lightning Source LLC
Chambersburg PA
CBHW080908170526
45158CB00008B/2036